LOST IN COOS

Other books by Lionel Youst

Above the Falls, 1992

She's Tricky Like Coyote, 1997

with William R. Seaburg, *Coquelle Thompson, Athabaskan Witness*, 2002

She's Tricky Like Coyote, (paper) 2002

Above the Falls, revised second edition, 2003

Sawdust in the Western Woods, 2009

Cover photo, Army C-46D aircraft crashed near Pheasant Creek, Douglas County – above the Golden and Silver Falls, Coos County, November 26, 1945. Photo furnished by Alice Allen. Colorized at South Coast Printing, Coos Bay. Full story in Chapter 4, pp 35-57.

Quoted phrase in the subtitle is from the subtitle of *Pioneer History of Coos and Curry Counties*, by Orville Dodge (Salem, OR: Capital Printing Co., 1898).

LOST IN COOS

"Heroic Deeds and Thrilling Adventures"
of Searches and Rescues on Coos River, Coos County, Oregon 1871 to 2000

by Lionel Youst

Including material by Ondine Eaton, Sharren Dalke, and by Simon Bolivar Cathcart

Golden Falls Publishing Allegany, Oregon

Golden Falls Publishing, Allegany, Oregon

Printed in the United States of America
ISBN 0-9726226-3-2 (pbk)
second imprint, 2019

Frontier and Pioneer Life – Oregon – Coos County – Douglas County
Wilderness Survival, case studies

Library of Congress cataloging data
HV6762

Dewy Decimal cataloging data
363

Youst, Lionel D., 1934 -
Lost in Coos
Includes index, maps, bibliography, & photographs

To contact the publisher

Lionel Youst
12445 Hwy 241
Coos Bay, OR 97420
www.youst.com
lionel@youst.com

To Desmond and Everett

To preserve from decay the remembrance of what men have done

– Herodotus

Table of Contents

Maps

Note: All the maps were drawn by the author using USGS maps as base, except map I, which is from a U. S. Fish and Wildlife map and Map VII, which was drawn by Jerry Phillips from an Elliott State Forest map.

Acknowledgments

First I have to thank Ondine Eaton because back in 1977 when she was a little girl of seven she was lost for three days and nights a few miles above our house on the East Fork of the Millicoma River. In 2009 I wanted to tell her story and she graciously gave me her account of it. It was so good that it inspired me to go ahead and research and record the rest of the stories that appear in this anthology.

Next, I must thank my very good friend Nathan Douthit, professor of history emeritus at Southwestern Oregon Community College. He read most of these accounts in their early drafts and was generous in sharing his very considerable critical skills with me. I took all of his suggestions very seriously and made many changes as a result. The deficiencies that remain are strictly my own, because I didn't do everything he suggested.

Next I thank Gordon Ross. Very early after my decision to put this together, he introduced me to the story of Beasley and Perdue, which became the first chapter in the book. Without it, there would have been nothing from the pioneer period. Then, I have to thank Dennis and Lynne Rice who helped me with two of the stories – the one of Beasley and Perdue because it was preserved through the efforts of Dennis's uncle Dr. Phil Morgan; and the story of the rescue of the lost hunter by Harold Noah who, by a fine coincidence, was father of Dennis Rice's sister-in-law. Lynne's genealogical research helped to keep me from several serious errors of fact.

Jerry Philips, a long time friend who has helped keep me straight for many years on matters concerning the Elliott State Forest helped me again. On matters concerning the Elliott I have also to thank Allen P. Krenz and Greg Kreimeyer. Others with great knowledge of the forests who provided information were Ralph Sweet, Scott Starky and Frank Lyon.

Wally Noah gave me invaluable leads in my search for the story of Harold Noah and the rescue of the elk hunter, Clayton Carroll. The most important of his leads was Jimmy Leeth, who in turn led me to others. Clayton Carroll's daughter Patricia Wallin, the last living person with acqaintance of the search and rescue of her father, was extremely helpful. On the story of John Clarence Fish, I am indebted to information from his daughters Kathleen Moore and Candace Pressnall. For the tragedy of Harold and Erma Ott I am indebted to their son Roger. And I am indebted to Melissa Green for important information on the death of her great-uncle, the trapper Frank Bremer. Sharren Dalke and three of her rescuers, John Saxton, Mike Johnson, and John Entgelmeier, made it possible to tell her story.

I am indebted to others, too numerous to mention, but I cannot omit my good friend Al Lively, who grew up at his grandfather's place on the West Fork and filled in many historical facts. My two sons-in-law, Larry Otten and Mike MacRae helped, each in his own way, as did my daughters Alice and Julia – and Julia, as usual, copyedited most of it. My son Oliver, whose five years in the 10th Mountain Division taught him a lot about survival, gave me a knowledge source I could tap when I needed it.

Vicki Wiese at the Coos County Historical and Maritime Museum, and Carole Ventgen at the Coos Bay City Library, were always there to help with specific research problems.

And most important, I have to acknowledge my wife Hilda, who for 50 years has been with me on whatever I want to do, and I love her for it.

Map I

Northern Coos County and part of Western Douglas Counties
Showing major forest ownerships
(Note area of the cites, North Bend and Coos Bay, in relation to the forest area)

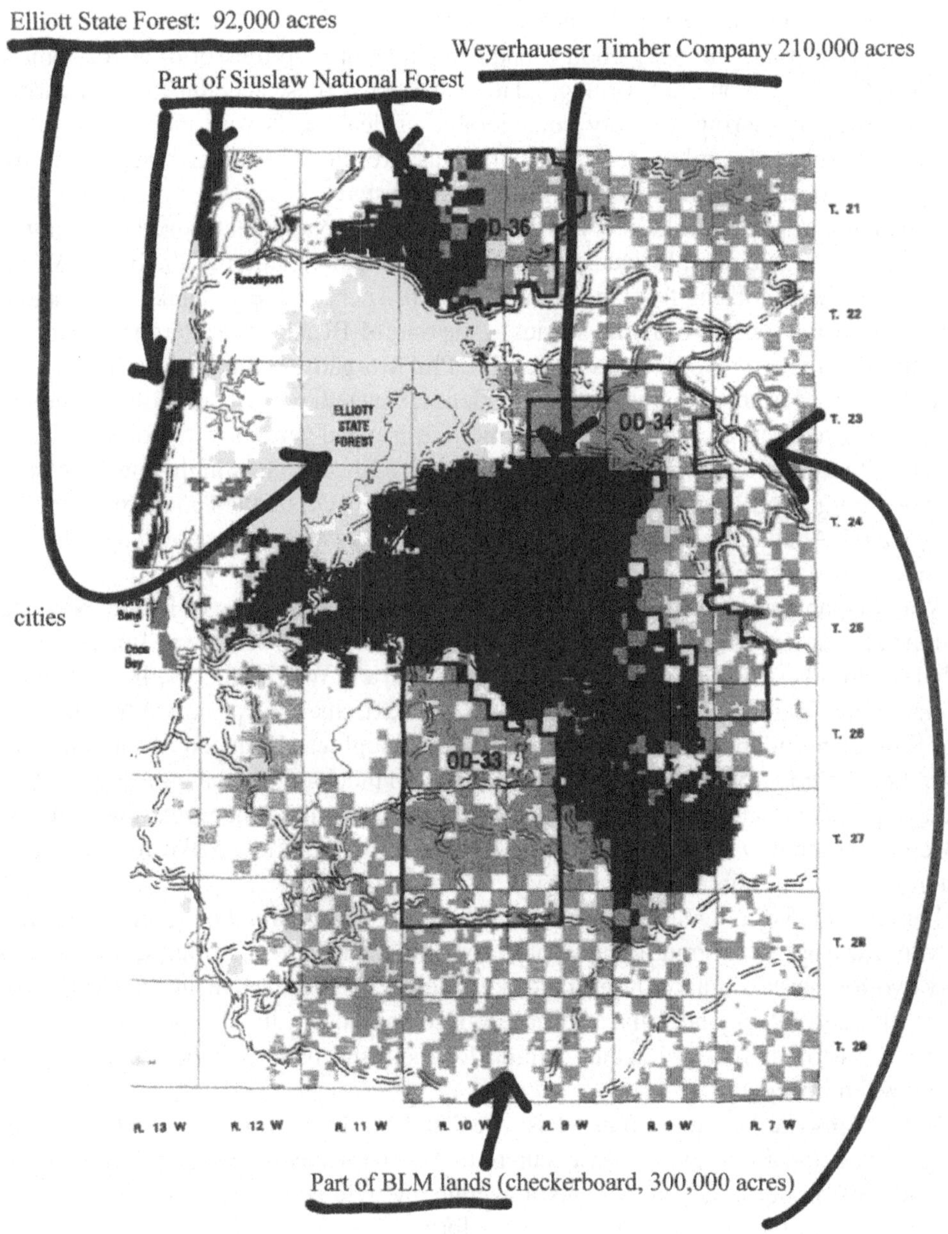

Introduction

My first goal in writing this book was to tell the stories of persons lost in the woods of the Coos River drainage of Southwestern Oregon. It became more an anthology than a unified work because as it grew, it took on incidents of search, and sometimes of rescues, some of which were barely in the woods. My good friend Nathan Douthit made light of it after he read Chapter 3 "Splash Dams" and Chapter 9 "Road Slides." He suggested that my title and subtitle might better be "Lost in Coos: Stories about People Lost and Sometimes Found or Searched for and Sometimes Rescued; as well as Other Stories of Fatal and Near Fatal Accidents, Misfortune and Good Luck." Although tongue in cheek, as usual he is quite right. And I make no apologies. I live in these woods and, as various as the stories are, they fill in some of the otherwise unrecorded history of the 85% of northern Coos County that almost nobody lives in, the woods. I think that is something worth doing.

My interest in the subject probably goes back to November 1945, when I was twelve years old and an Army C-46 aircraft had crashed into the wilderness area someplace above the Golden and Silver Falls (see Chapter 4). It was a major topic of all conversation for a month and my brother-in-law, Jerry Baughman, was on the search from the beginning. When it was over, he brought me one of the propellers from the plane, which I kept in my room until I left home. Then, there was the search for Glenn Thornton, who was lost in a log drive on the South Fork in June of that same year (see Chapter 3). He was step-brother to our neighbor and elementary school chum, Albert Lundberg. Later, the story of Harold Noah and his daring rescue of the elk hunter Clayton Carroll in 1950 held my attention (see Chapter 5). I was personally involved in a couple of the searches, the one for John Clarence Fish (see Chapter 8) and the one for Ondine Eaton (see Chapter 7). And it was my attempt to re-tell the story of Ondine Eaton that stimulated my interest in searching out and resurrecting the collection of stories that follow.

These are stories that I fear have been lost in the current generation and part of my self-imposed task has been to bring the reader along with me as I retrieve some of them. They derive partly through oral and folk traditions (a long-time interest of mine), partly through newspapers and other written accounts, partly from my interviews with living participants in the events, and partly from my own personal experience. I have used whatever sources I could find and it was the chance conversation that frequently led me to some of my best and most surprising of them.

I think the stories are worth preserving because, among other reasons, they have relevance for local history. They reflect changes, over time, of the relationship of the settled parts of the area (about 15%) with the forested parts (about 85%). With populations encroaching ever more closely into forested and wilderness areas wherever they exist, these stories may have a more general and universal appeal. I confess, however, that what I really want is to tell some engaging stories that will hold interest. If I've done that, I have succeeded.

The stories arrange themselves quite naturally into three periods of the local history. The first one took place in 1871, before the land was fully explored or surveyed. It properly belongs by itself and falls into the "Pioneer Period." The next group of stories – 1938, 1945, and 1950's – are mid-twentieth century, after the land was surveyed but before major road building and logging had begun. The lost persons in these first two periods (Part One of the book) are all men, and none of them are vehicle related. The last set of stories (Part Two of the book) are from after the Columbus Day Storm of 1962 and are characterized by the existence of hundreds of miles of logging roads throughout both the Weyerhaeuser and the State Forest.[1] All but one of the stories in this last set are vehicle related, and the object of most of them have been women and children, and among the adults, the elderly.

The scope of the book is limited to the Coos River drainage of Coos County, Oregon. One obvious reason for such a limitation is that I have lived there most of my life and know quite a bit about it. There is a wonderful variety of kinds of searches, and of rescues, and collectively they paint a picture of the human uses of this vast forestland throughout the historic period. Over the years these woods have been penetrated for exploration, hunting, trapping, homesteading, collecting minor forest products, surveying, cruising timber, logging, and after the roads were built people drove their cars up there and some of them got lost, and some of them died. There were among the searches and the rescues some extraordinarily heroic deeds. Some of the stories are good enough that they need to be preserved for their own sake.

There may be a more specialized interest on the part of certain readers. Until fairly recently, searches had a somewhat random or arbitrary character about them. There was apparently no specific criteria as to when, or whether, to begin or end a formal search, nor of who or what agency was responsible for carrying it on (although it was usually, but not always, assumed to be a sheriff's job). These stories demonstrate, however, that once it was known that there were lost persons in the woods, there was never a shortage of

---

[1] Most of the original surveys were conducted from 1889 to about 1910. Immediately following a survey, the Northern Pacific and the Oregon and California Railroads hired timber cruisers to locate the timberlands that would be most attractive for quick sale to the lumber companies that had cut out their holdings in the Great Lakes and were looking to the Northwest. In the Coos River drainage it was Weyerhaeuser and Pillsbury that purchased most of it, although Waterford Lumber Company, Menasha Woodenware, and a few others also got in on the action. In the Coquille River drainage of southern Coos County, the C. A. Smith company of Minnesota purchased the largest tracts. The lands of the Coos Bay fire of 1868 were of no interest to the railroads or the lumber companies and were absorbed into the Siuslaw National Forest and later turned over to the State of Oregon to form the Elliott State Forest.

volunteers willing to sacrifice their time and in certain cases to risk their lives to save the lost persons.

When we hear of search and rescue in the mountains or in the woods, it often involves lost hikers, climbers, hunters and other users of wilderness or unpopulated mountainous recreational areas. The forests of northern Coos County, however, are considered as commerical or industrial forests (although parts of them are quite wild). Persons have sometimes become hopelessly lost in them, but the lost persons have seldom been of the recreational type we hear of in the wilderness areas.

Every location on earth has its own local peculiarities and the Coos River drainage of Coos County, Oregon is no exception. The area comprises about one-third of the county, some 960 square miles more or less, all of it on the western slope of the Coast Range. The terrain is extremely broken with steep, sometimes vertical slopes, and frequent sandstone bluffs. Most ridges are from 800 to 2000 feet above sea level and the canyons are usually choked with brush and windfalls. The rivers and streams are fast and crooked, the longest (South Fork, Coos River) being 63 miles. About 85% of the area is uninhabited forestland, a temperate rainforest which in nature usually develops through a succession following catastrophic forest fires. The period covered is from 1871 to 2000, and the principal fires occurred in about 1660, 1770, 1840, and 1868, providing at least four distinct forests as we come into the local historic period. There were many other lesser fires over the years, including a conjectured fire of 1440, giving the area considerable variation of age and species but pure stands of even-age Douglas fir tend to dominate. Douglas fir is one of the world's most important structural woods and its commercial importance has driven the man-made changes in the forest to the present day.[2]

Until 1950, most of the logging within the Coos River drainage was done on lands that were around the periphery of the two forests that are within the scope of this book. Some of that logging was by rail and some on the rivers using logging splash dams.[3] Road

---

[2] Western hemlock and western red cedar is often found as part of the understory in the late succession (and sometimes in pure stands), with patches of red alder and bigleaf maple usually in the early succession. Bare spots, free of trees, were very rare except in places of more recent fires. I am grateful to Jerry Phillips for providing me a map in which he has drawn the boundaries of many of the historic fires. An even-aged Douglas fir forest frequently emerges after a catastrophic forest fire and it is a simple matter to determine the date of the precipitating fire by merely counting the annual growth rings of the trees – although successive re-burns within part of a previous burn are not uncommon and tend to complicate the matter.

[3] We are most fortunate in having two books that document the splash dam and the railroad logging in Coos County. See Dow Beckham, *Swift Flows the River: Log Driving in Oregon*, Coos Bay, 1990; and William A. Lansing, *Can't You Hear the Whistle Blowin': Logs, Lignite, and Locomotives in Coos County, Oregon*, 2007.

building for truck logging did not begin on the Weyerhaeuser Millicoma timberlands until 1948 and the first logs were not taken out until 1950. On the South Fork, Weyerhaeuser road building did not begin until 1962, a pivotal year for these woods and these stories because that is also the year of the Columbus Day Storm. The massive amounts of timber blown down in that storm brought a glut onto the domestic lumber market and opened an entirely new market in Japan. This in turn brought about accelerated logging and a very rapid liquidation of the Weyerhaeuser old growth, and it ushered in a new kind of lost person – the ones that are "vehicle related." Once people from town could drive their cars up into those woods, the lost persons soon became of a type quite different from the lost persons of the days before the Columbus Day Storm.

What do I mean by "lost?" I use the common approach that if somebody is looking for them, they are lost. They may not think of themselves as lost, but they are lost to somebody. The nineteenth century was different. Frontiersmen of that period tended to find their own way out. In southern Coos County in 1851 there was a case in point: the T'Vault party, trying to make a trail from the coast at Port Orford to the gold diggings at Jacksonville in southern Oregon. They came over the ridge between the Rogue River and the South Fork of the Coquille River thinking they were on the Umpqua. Hopelessly lost, struggling down the South Fork, they crossed over another ridge, this one from the South Fork onto the Middle Fork of the Coquille, and down it to friendly Upper Coquille Indians who gave them canoe transportation down the main stream of the Coquille where they were attacked by unfriendly Lower Coquille River Indians near the mouth of the river. Five of the party were killed. One escaped to the south and two to the north, one seriously wounded. By the time the two who headed north arrived at help near the mouth of the Umqua, they had been gone from Port Orford for a month and were barely alive.[4]

I consider the T'Vault expedition outside the scope of this book because it was on the Coquille River drainage and I am limiting these stories to the Coos drainage. The earliest case of a party being seriously lost that I am aware of in the Coos drainage was the hunters Beasley and Perdue in 1871. Their case constitutes the first chapter in this book.

---

[4] William G. T'Vault, a promoter of Jacksonville and the Rogue River Valley, led this ill-fated expedition. See Stephen Dow Beckham, *Requiem for a People*, University of Oklahoma Press, 1971, pp 59-63, Nathan Douthit *Uncertain Encounters*, pp. 117-8, and Youst and Seaburg, *Coquelle Thompson,* p. 21-25 for accounts of the T'Vault expedition.

PART ONE: Before the Roads

Part one consists of the first six chapters, all of which cover incidents that occurred prior to the Columbus Day Storm of 1962. The major construction of logging roads had not yet begun, and all of these searches and rescues took place in a comparatively roadless area. Coincidentally, all the lost person in Part One are men, and the persons doing the searches are also all men. Chapter One is set in the Pioneer period, before the land had been fully explored. Chapter Six is of an airplane and its pilot lost before the Columbus Day Storm but not found until well after it.

Map II

Coos River Drainage, Northern Coos County, Oregon

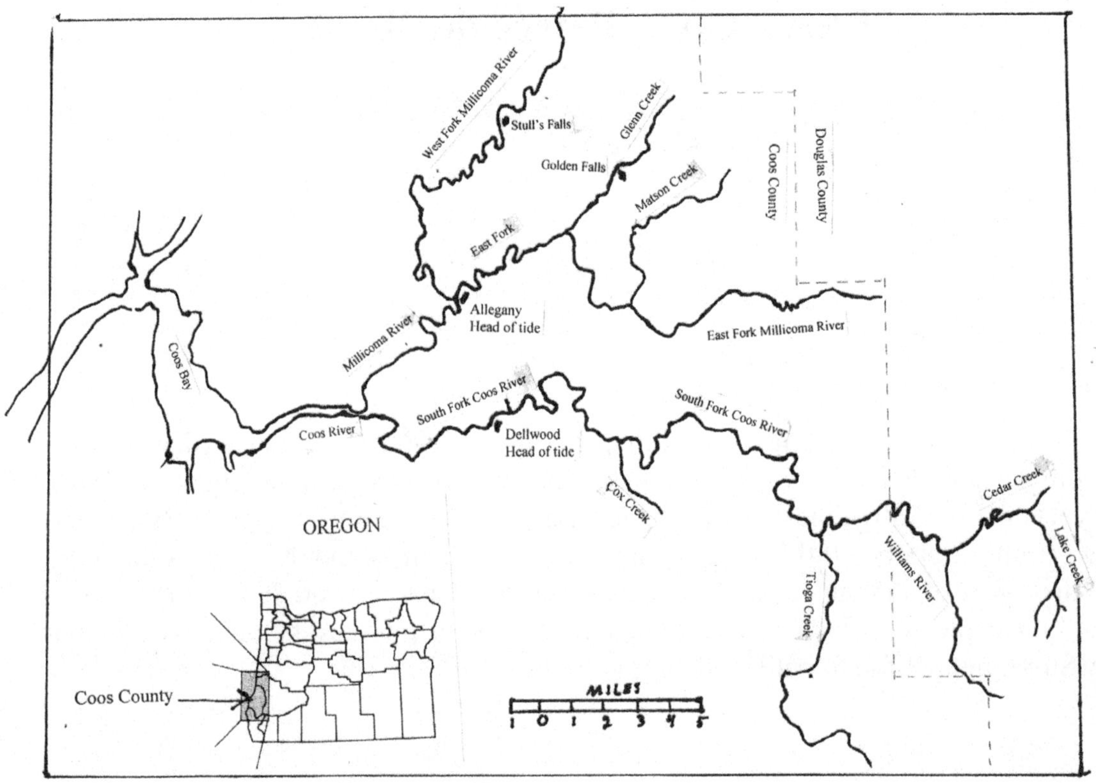

Chapter 1

Beasley and Perdue

Two hunters nearly die of starvation and exposure in 1871

S. B. Cathcart, Civil War veteran and Coos Pioneer, laid to rest with military honors.

"The Lost Gun"
by
Simon Bolivar Cathcart[1]

In March 1871 John Perdue and John Beasley went up on the Coast Range mountain west of Flournoy Valley [Douglas County] for an afternoon hunt. They traveled north along a ridge which they thought was the summit, but was only a spur leading west into the headwaters of Coos River. They could not tell anything about the course they were traveling on account of the rain and fog.

After they had gone far enough to be as they supposed to be about west of Perdue's home at the foot of the mountain, they turned off to the right down the mountainside and found a small stream which they supposed led down to Perdue's house, but which was the headwaters of Coos River. They followed this down until dark

---

[1] Simon Bolivar "Dal" Cathcart (1842-1932). Born in Indiana, moved to Oregon with his parents in 1853. Co. A, 1st Oregon Cavalry during Civil War, after which he taught himself surveying. In March, 1871 he led a party in search of two hunters lost near the divide of the Coast Range and this is his account of it, written by him on August 26, 1926. Settled at Allegany, Coos County in 1871. 1873 appointed US Deputy Mineral Surveyor of Coos and Curry Counties. Coos County Surveyor for 22 of the 32 years between 1874 to 1908. Active in the GAR, and he was the First Grand Chancellor for the Knights of Pythias of Oregon. Photo, above, is from the obit in the *Marshfield Sun*, May 19, 1932.

I copied the account from a singlespaced seven-page typescript provided me by Gordon Ross, and from copies of original manuscript pages provided me by Dennis and Lynne Rice. I did very light editing, primarily by paragraphing and correcting obvious typographical errors in the typescript. In a few cases I changed punctuation for clarity, and in one case I moved a paragraph from near the end of the story to its chronological place in the narrative. Other than those few changes (and the footnotes I have inserted), the story is as I received it in typescript. LY

and then tried to build a fire but their matches were wet and they failed to get any fire, so they sat there on a large flat stone all night. Their clothes being wet, they suffered greatly from cold.

The next day they followed down this little stream for several hours, when they were convinced that they were on the wrong side of the mountain. They turned to the right and ascended a high ridge, but could see nothing on account of the rain and fog. They descended the other side of the ridge but found another stream that was also a tributary of the Coos river. They followed this stream down until night when they lay down in the rain.

The next morning they traveled down the stream believing that they were on the water of Brunnit Creek, and that they would come to Brewster Valley. At night they again lay down in the rain without food and were suffering greatly from cold, being wet for three days and nights without any fire.

The next day they resumed their travel down the stream. About noon they came upon a small herd of elk but their guns, being old- fashioned percussion caps, were wet and failed to fire and they were left without food again.[2] Although the elk stood and looked at them until they were convinced that their guns would not fire.

Soon after this they found a flint rock and taking a small piece of cotton from the padding of one of their vests, they managed by striking the flint with a knife to start a fire but they being very weak from exposure and hunger, and everything being wet, they could build but a small fire. They succeded in drying their guns, and went back to where they left the band of elk, but they were gone. They took some cotton from their vests and slightly moistened it and worked some powder in it and carried it under their arms to keep it dry.[3] So with their flint rock and knife they could start a fire any time, but the only place they could be sure of getting fire was in some rotten stump where there was dry rotten punky wood.

That day they came to a large creek coming in from the right. When Perdue was satisfied that it was larger than Brummit Creek and that they were evidently on Coos

---

[2] The gun was a "Kentucky" style percussion lock musket with a round 40-inch barrel. See photo, page 13.

[3] According to an earlier account given by Cathcart (*Coos Bay Times* Dec. 30, 1924, p. 6), "they made lint of their driest cotton underwear and after a long time started a fire. After this they dried their powder and mixed some of it with the lint and had no further difficulty in starting a fire any time."

River, the only course then was to head down stream in hope of reaching settlement.[4] The only food they were able to obtain was frogs and snails.

The Rescue

On Friday about the 17th of March [1871], I was going on horseback from Coles Valley to French settlement (now known as Melrose), when I met Dr. Royal who halted me and says, "Dal you are wanted." "What is the matter, Doc?" I said; and he replied that John Perdue and John Beasley were lost in the mountains since last Saturday, out in all this storm, and all the able-bodied woodsmen are wanted for the search. I hastened on to French settlement[5] and found there were a number of men gathering there ready to start the next day to see if they could get any trace of the lost men. The next day we started about 25 strong made up of old men and boys, not more than 4 or 5 that were fit for an expedition of the kind.

We crossed the summit and camped overnight. In the morning R. A. Woodruff and myself and one or two others talked the matter over and all agreed that it would not do to go ahead with that crowd as we would soon have the woods full of lost men. So we all went back to the valley. I got my horse and went to Flournoys where a man by the name of Collins lived, and found only one or two men there. Collins said he would try and find a few more by morning so I went up to Winstons on Sugar Pine Mountain and stayed all night and in the morning went back to Collins and found 5 or 6 men, some of them good mountain men. (The reason I write so much about getting good men is I realized that to go into the mountain in that storm and trace up those lost men who had 9 or 10 days the start in wandering through those mountains, and get them out was no boy's job: it would take good men with plenty of nerve).

---

[4] Believing that they have to press ahead as their only course of action is all too common in the behavior of lost persons. Analysis of lost person behaviors reveals that lost persons almost never retrace their steps. In this case it came very near to being a fatal mistake.

[5] Near present day Melrose, Douglas County, the "French Settlement" was near the rendevous point for Alexander McLeod of the Hudson's Bay Company at the end of his expedition into the Coquille River Valley, October, 1826 to February, 1827. McLeod had come over the divide of the Coast Range by way of the North Fork of the Coquille River. As far as is known, no one had ever gone the length of the South Fork of Coos River prior to 1871.

While discussing the matter we saw a delegation of men coming from Coles Valley and Calaponia, among whom were George Sacry, Billy McBee, Tom Blaine, Alex and Henry Churchill, Ben Titmus and others. I told Collins that that crowd was good for any emergency. They had plenty of provisions for a number of days, a gun and two axes, no blanket to encumber them. In the crowd were three pocket compasses and a bugle. I suggested that we cross the summit and camp and in the morning that we place the bugle in the middle with one compass and then string out with north and south, keeping in hailing distance with the two other compasses on the wings and all travel west looking for tracks of the lost men.[6]

This was agreed to but was found unnecessary for when we got down to the small stream where we were going to camp, we found a notice that was signed by Robert Kincade, father-in-law of Perdue, and Huntley, and Frank Conn. Both of the latter were perfectly unfit for the trip, so they had gone back for reinforcements and provisions and was coming in the next day. They had gone up a different spur from the one we came down and we had missed them.

We immediately struck camp and prepared a big log fire for by this time it had begun snowing. By morning the snow was 1 inch deep but melting rapidly. There were 13 of us altogether, so we deemed five would be a sufficient crew. So five of us took all the provisions that the 13 had, and the 8 went back. The five that constituted the rescue party were George Sacry, Billy Mc[G]ee, Tom Blaine, Alex Churchill and myself.

The snow soon melted so we could find the lost men's tracks in the elk trails. After traveling down the stream some time we lost their trail, but found it again at the junction of the stream with the stream before mentioned, which they had found when they crossed the ridge the 2nd day after being lost.

We traveled until 3 o'clock when we camped. We cut down a cedar and split out shakes about 10 feet long and about a foot wide, using our axes and wedges to split them out. They made a very good roof, so we had a fairly comfortable night. The next morning at daybreak we were off following down the stream. About noon we found a small band of elk and Sacry killed a two year old cow. We partly skinned her and cut out 6 or 8 pounds of meat for each one and put in our packs.

---

[6] Cathcart's proposed use of the bugle foreshadows a recent advance in ground search technique, the "Sound Sweep Search." See Coos County Sheriff.com, "Elements of a Sound Sweep Search," by Martin Calwell, 1998.

A mile or so farther down we came to a large creek (Cedar Creek), coming in from the Northeast. As there were no trees near that would reach across we had to go up the creek about 1/4 mile to find a log to cross on. Just as we got back to the mouth of the creek we heard a shot. We answered it and heard a Whoop, and soon saw Bob and George Woodruff coming in a run followed closely by Robert Kincade and W[illiam] W. Cathcart. By falling the older trees from each side of the stream they were able to cross without going up as we had done. They had replenished their meat supply from the elk we had killed. As it was still warm, they knew they were getting nearly up to us.

At 3 o'clock we halted and cut down a fir tree and made a good fire and after cooking and eating a good supper we laid down in the rain for the night. The rain fell in torrents all night. We had no shelter, there were 3 single blankets in the crowd, but we had a roaring fire and we could keep warm but some of the men could not sleep with the rain beating in their face. At daylight we were off and traveled all day with all the strength we could muster. It was a hard day as the river runs through canyons most of the way. The elk trails would cross the river wherever the bluff rock would come to the river, so we would have to leave the trail and climb steep rocky hills for hundreds of feet to get over the bluff and then descend again to the river bottom only in a short distance to have to climb over another bluff.

At night, weary from our strenuous climbing, we lay down in the rain, which never ceased for an hour. The next day (Friday) was about the same, although not so many bluffs to climb. That afternoon we found Perdue's gun sitting up against a tree. Knowing that Beasley was the younger and stronger man, we felt encouraged for as we had not found his gun, and that Perdue had been able to carry his gun this far, that they would be able to make it through.

But a little later in the day we were badly discouraged. We came to a big tree that had fallen down, and the top in the river. We found Beasley had not been able to make it and went back to the bank where the log was above the sand and dug the sand out so he could crawl under.[7] We could tell their tracks apart, Perdue wore tight high

---

[7] This needs elucidation. The search party had not come upon the lost persons, but had examined their tracks and from the evidence had found that Perdue, the weaker of the two, had left his gun but he was able to crawl over the log. Beasley, the stronger, could not get over the log and had to crawl under it. In his 1924 interview with the *Coos Bay Times* Cathcart expands: "We afterwards learned that earlier in the day they had built a fire in the roots of a stump which had fallen over on Beasley, injuring him greatly." (Dec. 30, 1924, p. 6)

heeled boots. Beasley wore wide loose shoes. Any 5 or 6 year old child could have climbed over the log. We were satisfied that he was all in.

Friday night the rain was not so heavy but it was still disagreeable. Saturday morning we were off at daylight and about 9 am we came up to them where they had a little fire in the root of a tree, and were hovering over it. They had given up and decided to die there.[8] They could scarcely speak above a whisper. They had a few snails and a frog spitted up in the bark of the tree. Immediately six of the men threw off what little packs they had and said they would try and make it through to the settlement, some to take the news back home and some to bring us food. All the food we had left was about one pound of elk meat, and an ounce or two of bacon and two or three pounds of bread crushed into crumbs. We had plenty of sugar, salt and coffee.

I immediately started a fire and made a little broth of elk meat and bacon, but their appetite was so far gone that they could eat but a spoon full or two. We carried them down to the river where we found a small sand bar with some drift wood and soon had a good fire. We got some boughs and ferns which we placed on the sand and spread a blanket over it which answered for a bed.

We pulled Beasley's shoes, which were large and loose and found his feet in good condition, but Perdue had on tight fitting boots and we had to split them to get them off and his socks were literally welded to his feet. We had to rip them open and soak them with warm water to remove them from his feet, which we found in a horrible condition.[9]

Here the sun shone out for a little while, the first time we had seen it since we started. As this was a hard place to camp, Sacry and I went down the river about one fourth mile and found a fine place to make camp. It was a fine level bench where since then the C. C. Cox homestead cabin stood. On our way back I saw a fresh deer track, so

---

[8] The men were about eight miles above the head of tide on the South Fork, on the north side of the river across from the mouth of Cox Creek. It had been fourteen days since they had gone off on an afternoon's hunt. They had traversed more than 40 miles through the brush and over rock bluffs, down an unexplored river in the rain, without food or shelter. They would not have survived exept for the rescuing party.

[9] Cathcart elaborates in his 1924 newspaper interview: "His feet were covered with running sores that did not heal for several years." (*Coos Bay Times*, Dec. 30, 1924, p. 6) He was suffering from what is now called Immersion Foot Syndrome (Trench Foot), a sometimes fatal result of wearing tight-fitting, non-breathing boots for extended periods.

after we had got Perdue and Beasley down, and got started to making camp, I took Sacry's gun (an old-fashioned Winchester) and started up the mountain to see if I could get some meat.[10] After traveling about half a mile, I saw a deer get up in the brake and ferns and started to walk away. I raised my gun and fired quickly and the deer fell. I was so anxious to get the deer that I was going to take no chances of its getting away that I threw another shell in the gun without taking it from my shoulder, and kept it pointed ready to shoot if the deer got up. After waiting a few minutes, I went step by step with the gun to my shoulder, ready to shoot until I got within a few feet of the deer. When I saw I had given her a dead shot, it would have been amusing if anyone could have seen me. I was like the boy that was after the wood-chuck – he said he had to catch it, they were out of meat. The deer was a fine large doe. I took out the entrails and cut the deer in two, threw the fore quarters up on a log and taking the hams astride my neck I started for camp. When I arrived it was just getting dusk. When I threw the hams down, Sacry made the woods ring. He said we were all right now if the boys were gone for a week. (We had no means of telling how far it was to settlement; whether the boys could make it; one day; two days or three or four days.)

The next day about 2 pm, R. A. Woodruff and Billy McBee accompanied by McKnight returned with some provisions but, as they had not waited long enough the morning before to see how bad the men were, they brought a limited amount of provisions with them as they expected to meet us half way out.[11] So I struck out for the remainder of the deer, which I brought and skinned and wrapped the skin around the head and neck and burried it under the coals and ashes of our log fire and left it there until about two o'clock next morning when I took it out and hung it up till next morning.

All these two days I had been giving the lost men some broth at frequent intervals but could rouse no appetite. In the morning, I took down the deer's head and the hair and skin was burned crisp, but when I peeled it off I found the meat cooked until

---

[10] There was only the one gun, Sacry's old Winchester. With someone competent to use it, as Cathcart obviously was, that was the only gun that was needed.

[11] William McKnight (1833-1898). From Virginia, arrived in Coos County in 1865 and settled at the head of tide on the South Fork where he lived for 32 years. The tributary to the South Fork of Coos River that the hunters and Cathcart's search party descended is called Williams River on the newer maps. I suspect that Williams River may be named after William McKnight.

it was tender enough to fall to pieces. I took out the tongue, and Beasley being awake, I cut a thin slice and putting a little salt on it, I gave it to him. He swallowed it and begged for more. At that time Perdue waked up and he too swallowed a little slice and he wanted more. Their appeitites were aroused, and they would cry and beg for food, but I knew that if I gave them all they wanted it would have killed them very quickly.[12]

I took the skin that was on the hams of the deer and sewed them, flesh side in, around Perdue's feet. We started out, half packing and leading the lost men, and in two days we reached McKnight's at the head of tidewater on Coos River. The second day McKnight met us with a horse which gave us quite a boost. The men were too weak to sit on the horse but by holding them on we could make better time. Their appetites were ravenous, and I had to feed them a very little at a time.

When we arrived at McKnight's, we were not very presentable, our clothes scarcely hanging on us. One of the men had a needle and a small amount of thread and I by accident had a small fish line in my coat pocket about [5?] feet long. After the thread had been used up in sewing up vents in our pants, we used the fish line very sparingly to tie up the worst tears and rips. We would punch holes in the cloth three or four inches apart and use just enough of the line to tie the edges together.[13]

We stayed all night at McKnight's and the next morning the others of the rescue party started for Douglas County, and I took Beasley and Perdue in a skiff and started for my brother-in-law's (John Bazzill) above where Allegany is now to keep them until they were able to be taken home on horseback. Perdue having learned the shoemaker's trade under Bazzill at Oakland in Douglas County felt perfectly at home there.

On their way out the rescue party met Beasley's father coming in after his son but Kincade, Perdue's father-in-law, turned him back. He told him that they would not be able to come for some time, but as soon as he got home his wife started another party in

---

[12] Cathcart recognized the danger of refeeding a victim of near starvation, and he did the right thing. The "Refeeding Syndrome" was not recognized clinically until the end of WW II when starving POW's were repatriated from the Pacific Theater. See, for example, "Much Ado About Refeeding," *Practical Gastroenterology*, January, 2005, p. 26ff. Also, "Refeeding Syndrome: A Literature Review," *Gastroenterology Research and Practice*, Volume 2011, Article ID-401971.

[13] This paragraph was at the end of Cathcart's manuscript. I have moved it to its chonological place in the narrative.

after him and they came near killing him. They had to lay at Sumner for 4 or 5 days before he was able to leave there. It seemed to be a hard matter to make some people understand that a person that was near starved to death, had to be treated as carefully as a typhoid convalescent. They seemed to think that all they needed was a good square meal. After keeping Perdue at Bazzell's for three weeks, I took him a row boat to Sumner – that is, where Sumner is located now – and got a horse from Charles Harned and took him to his home in Flournoy Valley.

I often think of this trip, of what we suffered, with no protection against the storm that raged continuously, and our traveling over those steep bluffs and canyons to the very limit of human endurance. Nothing would tempt one of the party, after they had gotten out, to undertake such a trip again, only to save human life. But when we would think of the fact that we had saved those men from a horrible death by starvation and exposure, and restored them to their families, we felt amply rewarded for all that we had endured.

As stated, we found Perdue's gun and brought it out with us, but Beasley's gun was not found and brought out until August 1925 when it was found by L. D. Smith and sons and brought out after laying there in the mountains for more than fifty-four years. I have it now in my possession and expect to turn it over to the Oregon Historical Society.[14]

It was so badly rusted and decayed that it would scarcely hold together. Both Perdue and Beasley and eight of the rescue party have passed to the Great Beyond and at this date August 26 1926 I am the only one left.[15]

---

[14] Following is from the files of the Oregon Historical Society, indicating receipt of the gun: Rifle; percussion lock. Owned by John Beasley. He, with John Perdue became lost while hunting in the coast range on the headwaters of Coos River, in March 1871. After several weeks search by a party of 9 persons, they were finally rescued, but nearly dead for want of food. One gun was found at the time of rescue. Later, this gun was found in August 19[2]5 by S.B. Cathcart and L.D. Smith. Presented by S.B. Cathcart, sole survivor of the rescuers, Marshfield, Oregon. September 1, 1925.

[15] In his interview with the *Times* in 1924, Cathcart continued: "Perdue died at Medford only a few years ago; Beasley died in Douglas county about eight years ago; Kincaid many years ago; Geo. Sacry died at Silverton about 1920; Thos. Blaine 10 years ago in Marshfield; Alexander Churchill at Oregon City in 1918; George Woodruff at Mercy Hospital, North Bend, in 1910; William McGee a few months ago at Seattle, and W. W. Cathcart last January in Modesto, Cal. (*Coos Bay Times*, Dec 30, 1924, p. 6.)

I am writing this from memory and although more than half a century has elapsed, every detail of the events is as fresh in my memory as if they were happenings of yesterday.

S. B. Cathcart

Simon Bolivar Cathcart was the first Grand Chancellor of the Knights of Pythias for Oregon. The fraternal order, formed in 1864, was inspired by a play based on the Greek legend of Damon and Pythias. It emphasizes trust and loyalty in true friendship, traits that appear in high relief during and after the search and rescue of Beasley and Perdue.
Photo from CCH&MM #996-P17

Beasley and Perdue

The once celebrated case of Beasley and Perdue is the earliest known search-and-rescue effort on the Coos River drainage of northern Coos County. John Beasley and John Perdue were hunters who in March 1871, got lost on the Umpqua–Coos divide west of the Flournoy Valley in Douglas County, and unknowingly came down into the headwaters of the South Fork of Coos River. Wrongly assuming that they had to continue downstream, forty miles and fourteen days later they were huddled under a tree waiting to die when they were found by a search party of nine men led by the redoubtable S. B. "Dal" Cathcart. It is a stirring and inspiring story, and contains many of the themes found in subsequent search-and-rescue attempts.

The story of the search and rescue of Beasley and Perdue is the only one we have of the nineteenth century pioneer period in the Coos River drainage. It occurred before the South Fork of the Coos River had been completely explored, and long before it was surveyed. As far as I know, this was the first anyone in historic times had traversed the entire length of the South Fork, from its source at the summit of the Coast Range to tide water. And there is some irony in the fact that when the area was finally surveyed during the 1890's, it was Cathcart who was the County Surveyor.

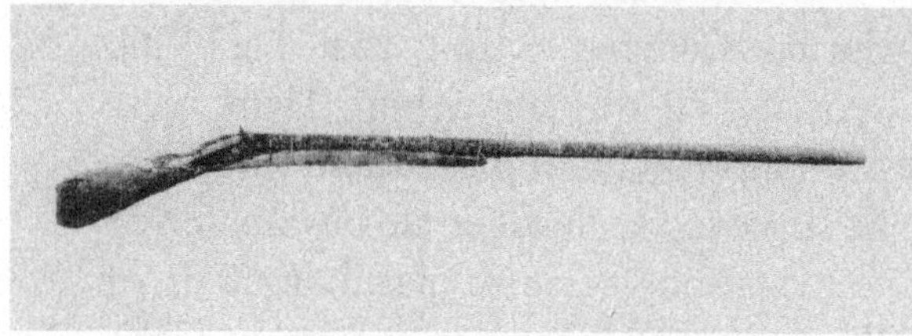

"The Lost Gun" is the title Cathcart gave to his narrative of the search and rescue of Beasely and Perdue. The gun, above, was a Kentucky style cap and ball muzzle loader with a 40 inch barrel. Picture provided by Lynne and Dennis Rice.

My interest in Dal Cathcart and his career is probably enhanced because my childhood home, and my current home of the past 35 years, is at Allegany near where Cathcart lived from 1870 to 1890. My own property lies between what is now the Elliott State Forest and the Weyerhaeuser Millicoma Forest, as did his. I counted 68 of the surveyor's field notebooks that Cathcart had filed in the Coos County Surveyor's office, and all of them were of original subdivision surveys of townships within the forests of the Coos River drainage from 1887 to 1898. Cathcart's legible handwriting and clear expository style grace every one of them. He is remembered in at least two geographic names in Coos County. Cathcart Mountain, at the head of Salmon Creek (in sections 2 and 3, T26S R11WWM) southeast of Dellwood, and Mt. Bolivar, the highest peak in the county. While the peak was officially named for Simon Bolivar, the liberator of

South America, it is no accident that Bolivar was also the county surveyor's middle name.

If there was ever a profession during Cathcart's time that had "job security," it was surveying. There were millions of acres of public land waiting to be surveyed so that it could be transferred to private ownership, and many tens of thousands of those acres were in Coos County. The County Surveyor was responsible to ensure that the surveys were done accurately and expeditiously, and from the evidence it appears that Cathcart did his job well.

I have included a greater degree of historical context than might be expected in a work of this kind, which brings us to the story of the provenance of the story: how we came to have it after 140 years. It was kept alive for the first couple of generations because many of the men who were on the search lived a long time and told it over and over. Most of them were from in and around the Flournoy Valley in Douglas County but some of them were residents of Coos County. Dal Cathcart had married Dora Landrith of Coos River in 1869, and they lived at the head of tide at Allegany. In March 1871, Dal was evidently visiting his brother William W. Cathcart, who lived in the Flourney Valley at the time, and who joined Dal in the search. George Woodruff and Thomas Blaine also joined the search, and they were brothers-in-law – Woodruff's sister Francis was married to Blaine, who had the ranch immediately below the forks of the river at Allegany. There were many close relationships among the rescued and the rescuers, who kept in lifelong touch with each other and who managed to keep the story alive as long as they were alive.

Cathcart, the leader of the search, was the exceptionally well known and popular Coos County Surveyor for many years. One of his friends was Lorenzo "Ren" Smith, who lived on Daniels Creek, tributary of the South Fork of Coos River.[16] In 1921 Smith's nephews, Nathan and Guy Cutlip, were running their traplines on the Williams River when they found the gun that had been left by John Beasley in March, 1871. It was in a crevice of rock on the north side of Williams River about about two miles upstream from where it joins Tioga Creek to form the South Fork of Coos River. Busy with their trapping, they left it where it was, but they knew the story of Beasley and Perdue and recognized that this was probably a gun left by one of them. They told about it when they got home later that winter, and when their uncle Ren Smith heard about it, he was intensely interested. He had lived on Daniels Creek in 1871 when the two lost men were found, and he was an old friend of Dal Cathcart, who was still living. The report of finding a tangible

[16] I am indebted to the account written by Dr. Phil Morgan, DVM (a grandson of Ren Smith), "Sequel to the Lost Gun," in which he details the events which follow. A typescript of Dr. Morgan's account was provided me by Gordon Ross, and also by Dennis and Lynne Rice.

artifact of that remarkable rescue of 51 years earlier stimulated the re-telling of the story, and generated interest in retrieving the gun.

The story began to circulate, and a reporter from the daily *Coos Bay Times* interviewed Dal Cathcart about it. The lengthy story appeared on page 6 of the December 30, 1924 editon of the paper, and it was very well told. The following summer 70-year-old Ren Smith decided to mount an expedition to retrieve the lost gun. He was well acquainted with the Tioga and Williams River area, having once filed and proved up on a timber claim in the Tioga Creek bottoms. For many years he had hunted elk there with others, drying and smoking the meat before bringing it out. From the description of the location of the gun, he thought he knew exactly where it was.

Lorenzo Dow "Ren" Smith (1855-1928) Photo © Lynne Rice, 2010.

He took two of his sons, Tom and Sam, along with their neighbor John Clinkenbeard, and headed up the South Fork. Ren Smith's age started to show up on him and he had to stop and camp near Fall Creek but he gave specific instructions to the younger men as to where they should find the gun. They found it with no difficulty and returned with it the next day. Sam Smith brought the gun to town, showing it to a reporter at the *Coos Bay Times* who wrote a somewhat garbled account, which appeared on page 8 of the August 27, 1925 edition. In it, however, is the best description of the gun. "It was an old Kentucky cap and ball. The rifle, a muzzle loader, has a forty inch barrel. At one time it was ornamented with brass but this has about disappeared."

S. B. Cathcart (1842-1932). Photo cropped from Civil War Veteran's group picture. CCHMM 988-P135

The reporter alleged that Mrs. Perdue, widow of the lost hunter John Perdue, wanted the old gun as a keepsake. When Dal Cathcart read that, along with several other errors in the article, he responded with a rebuttal, which appeared two days later in the competing daily newspaper, the *Southwestern Oregon Daily News.* Cathcart said that Mrs. Perdue had asked him to write a true statement of the rescue, inasmuch as he was the only living person who knew all the facts. A record of the Oregon Historical Society shows that on September 1, 1925, the gun was presented to them "by S. B. Cathcart, sole survivor of the rescuers." A year later, August 26, 1926, Cathcart penned the story that appears on the following pages. Len Smith's grandson, the veterinarian Dr. Phil Morgan, wrote an account of the finding of the gun and he saved the newspaper clippings and a picture of it. I have relied on his typescript in my retelling, and on the material in possession of Smith's great-grandson Dennis Rice who along with his wife Lynne, provided me with copies and for which I am very grateful.

It was, however, Gordon Ross who gave me the story in the first place and his account is a further example of how it was kept alive over many years. Gordon said that he first heard about it when he was 6 years old, "sitting with my grandfather on Catching Inlet; Tom Smith was the narrator." Tom Smith should have known the story very well indeed, as he was one of the young men who retrieved the gun in August, 1925.

The two main points that impressed young Gordon Ross was "of the two men trying to dig under a log, and the finding of the gun." Later, in 1965, Gordon again talked to Tom Smith about it. This time there was a further embellishment to the story. He said that when the boys returned with the gun to their father's camp near Fall Creek, "the first boy to come in sight was empty handed but the next one came in view pointing the gun at his father," meant as a surprise no doubt. Gordon said that his maternal grandfather, Henry Black, had started with the original search party in March, 1871, and was with the group that returned to Flournoy Valley the second day. He added that Thomas Blaine, who continued with Cathcart on the search, was "an orphan boy who had come west from Missouri with the Blacks." It was Thomas Blaine who later settled on the first ranch below Allegany (now owned by Ryan Mahaffy). There is a lot of local history in this story!

Tioga Dam, South Fork Coos River

South Fork, Coos River, a little below the confluence of Williams River and Tioga Creek. The Tioga Dam was built in 1944 to splash logs down the river, 22 miles to tide water. The picture show the rugged canyon of the South Fork which Beasley and Purdue, and their rescuers, followed. Photo was taken by Bill Wilt, about 1949 with a 4 X 5 Speed Graphic camera. It is used here with permission of the owner, Curtis Irish.

Map III

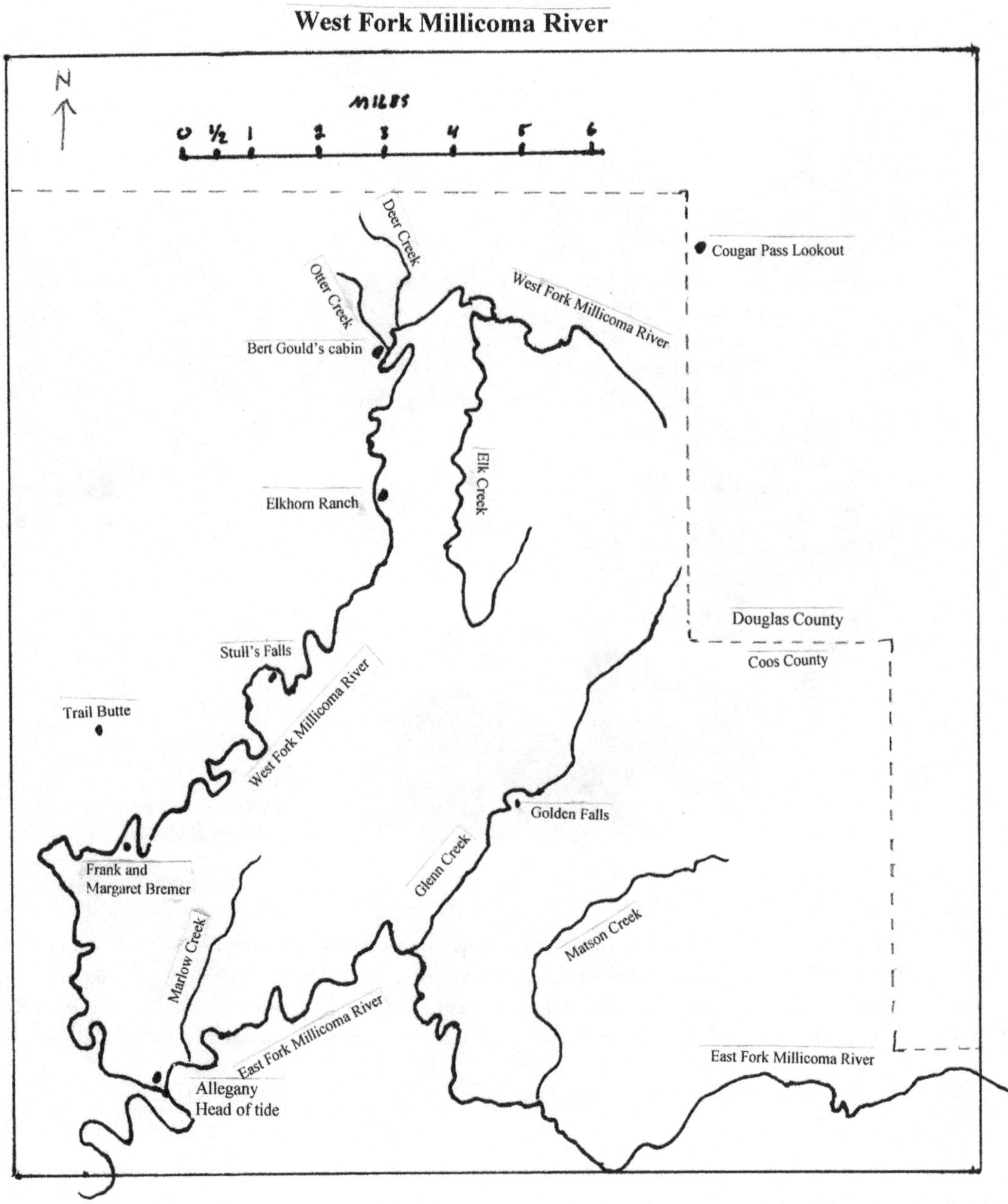
West Fork Millicoma River
N
MILES
0 ½ 1 2 3 4 5 6
Deer Creek
Otter Creek
Cougar Pass Lookout
West Fork Millicoma River
Bert Gould's cabin
Elk Creek
Elkhorn Ranch
Douglas County
Coos County
Stull's Falls
Trail Butte
West Fork Millicoma River
Golden Falls
Frank and
Margaret Bremer
Glenn Creek
Marlow Creek
Matson Creek
East Fork Millicoma River
East Fork Millicoma River
Allegany
Head of tide

Chapter 2

Frank Bremer Sets His Last Trap

The Elliott State Forest didn't even have an office at Coos Bay in 1938, the year that Frank Bremer "set his last trap," as Jesse Allen Luse of the *Marshfield Sun* put it. There had never yet been a timber sale on the Elliott, but there had been a recent three-year contract for peeling cascara bark, negotiated through the state office in Salem. In 1937 the contractor got out 20 tons of bark in three months, with an expected 75 tons by the end of the contract period.[1] That is a lot of cascara bark, but was the extent of the minor forest products officially harvested on the Elliott during the 1930's. There was no official policy for other products such as sword ferns, or trapping small furbearing animals. Those sorts of things were free for the taking in those days.

Trappers decided among themselves who would take one creek or another. Lem Gray had been on Elk Creek until 1936, coming in from above the falls on Glenn Creek where his wife Grace taught at the little Golden Falls School from 1934 to '36. Her $67.50 per month salary during the school year provided the steady income needed to keep a trapper financially afloat. When she finished her contract, Lem finished his trapping and they moved back to their permanent home, which was on the West Fork not far from Frank and Margaret Bremer.

Frank Bremer had been married for less than two years and had been trapping on the West Fork. This year he planned to expand into the Elk Creek territory vacated by Lem Gray. Mink and marten were the luxury pelts they were after, although a few dollars came in from the much lower valued racoon. A trapper could always hope to receive an occasional bounty from the county on a bobcat or lynx – worth $5. With the greatest of luck, he might even get a cougar and its $50 bounty.[2] That was a very rare event, but the hoped-for cash and prestige was enough to keep many a trapper trying year after year.

Frank Bremer and the other trappers were living near the very end of the long and colorful history of the fur trade. It was soon to take on a new phase, that of the "fur farm," but in 1938 it was still the individual trapper, with his lonely trapline out in the woods, that dominated the industry. Demand for furs of one species or another are driven by

[1] Jerry Phillips, *Caulk Shoes and Cheese Sandwiches* (1997), pp 109-111.

[2] Youst, *Above the Falls* (2003), pp 160-163.

the whims of fashion, and for a couple hundred years the fashion in men's hats in Europe called for felt made from the inner fur of the beaver pelt. The demand for those beaver pelts had been key in the original exploration and development of most of North America. It is not by accident that Oregon is the Beaver State, and until the fashion in men's hats changed in around 1850 (and the beaver were almost trapped out) no other furbearing animal compared in economic importance. Early in the twentieth century silver fox became fashionable for a few years, then beginning in the 1920's it was mink. In 1938, the year Frank Bremer set his last trap, there were 550,000 mink pelts marketed in the United States and 80% of them were from wild mink taken from lonely traplines throughout the nation.[3]

And so, what is a trapline? A trapline is a series of traps along a creek, or it may be said that it is the route along a series of traps. The term came into general use during the 1920's, when mink first became the primary luxury fur. Mink usually go along the sides of the creek and the traps would be set in blind spots along the bank. These are places where the mink has to stay close into the bank to avoid an obstacle such as a log, or deep water, a rock, or a stick. The steel, longspring #1 or #2 traps would be set at each place that looked like a good blind along the creek, and the trapline may extend for miles. Where possible, the traps are set in such a way that the mink will drown after being caught. After setting the traps, the trapper would then "run his trapline" a week later, harvesting those mink that were caught and re-set the traps. The trapping season ran from mid-November through December, the very time that the average settler in the woods had very little else to do. Mating is in January, and the young mink are born in the spring, and are on their own and vulnerable for trapping the following November.

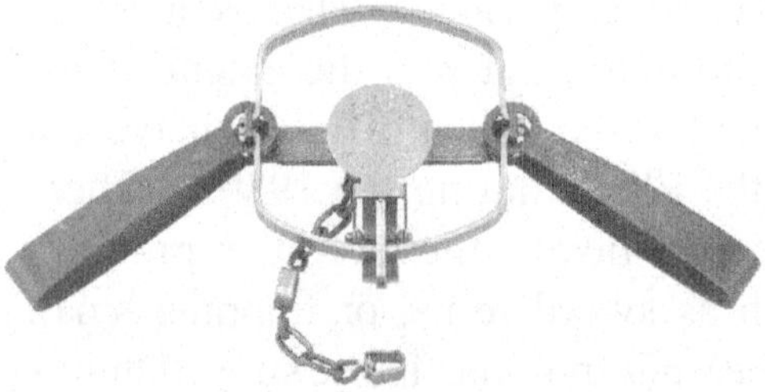

Victor Oneida #2 double longspring trap

On the Coos drainage, mink were trapped principally on a few remote and uninhabited creeks. Lake Creek[4] and Matson Creek, both on the adjoining Weyerhaeuser timberland, seemed to always have a trapper actively employed. Traplines were also run on the South

[3] Richard J. Aulerich, *Michigan's Fur Bearing Animal Industry*, Michigan State University (07/28/98), p.1. Aulerich notes that by 1970, 95% of the 5.2 milliion pelts marketed were farm raised.

[4] Lake Creek actually drains into the Umqua River, but was most readily accessible from Glenn Creek. It was part of the network of creeks this small group of Coos trappers exploited.

Fork in timberland owned by the Pillsbury family of Minnesota, on Tioga Creek and Williams River, but that was a world away and other trappers had that area sewed up. There was room for only one trapper per creek, and other than the state license, there were no other permits involved. Just move in and start trapping.[5]

A man named Lax had been working Matson Creek, but pulled his traps out and left for parts unknown. Cle Wilkinson had been on Lake Creek for several years. He waited a couple years for the furbearing population to rebound a little on Matson Creek, then he moved his traps over there. Gard Sawyer, from Elkton, moved his traps into Lake Creek. And now that Lem Gray was gone from Elk Creek, Frank Bremer was moving into that area.[6]

Over the years, hunters and trappers had built cabins throughout these woods. One trapper would move out and the next one moved in. Gard Sawyer moved into the Guerrin cabin on Lake Creek. Cle Wilkinson moved from there into Lax's cabin on the ridge between Glenn and Matson Creeks but he found it was too far from his trapline, so he built a new cabin of his own on Matson Creek. Frank Bremer was moving into the Bert Gould cabin on the West Fork. It was musical chairs. It took almost 500 square miles to provide enough mink and marten for three tappers!

Lem Gray, with lynx

Frank's father, Willian Bremer, had proved up on his homestead on the West Fork in 1892 and when he died in 1927, Frank and his sister Carolyn inherited it. Carolyn married Walter Stull, whose place was above them at the very end of the road on the West Fork, and in 1936 Frank married Margaret Browning. Like Lem Gray's wife Grace, Frank Bremer's wife Margaret was also a school teacher. She had graduated from the University of Nevada in 1923 and taught school in Nevada before the family moved to Allegany where she later served as principal of the Allegany

[5] Alice Allen informs me that her father, Cle Wilkinson, also trapped marten at a ratio of about one marten per six mink. The marten were trapped along the ridges, not the creeks. They were about the same size, but more slender than mink, a "lovely chocolate brown with a white spot on their throat," according to Alice. They spent much of their time in the Douglas fir along the ridges, and sometimes had large splotches of pitch on their pelt, which had to be cleaned before the pelt could be sold. Generally, the marten pelt brought a better price than mink. (interview with Alice Allen Sept. 19, 2010). At present, the marten is virtually extinct in Oregon.

[6] Youst, op cit

School.[7] She and Frank had been married less than two years and had no children when the 1938 trapping season started.Next above Stull's and accessible only by packtrail was the Elkhorn Ranch, established in 1885 by George Gould.[8] George made most of his living by selling elk products: meat, hides, trophy heads, elk teeth, whatever would sell, until the elk were almost extinct. In 1910 the State placed a moritorium on hunting them, a moritorium which lasted until 1940. The Goulds moved out after the hunting ban went into effect, but Clarence Gould, who was raised on the Elkhorn from the age of two, continued to use the area seasonally and maintained his hunting cabin on Deer Creek. Clarence Gould was married to Lem Gray's sister Jesse, so almost everybody up there was related to each other in one way or another. It was one of the Gould's cabins that Frank Bremer would be moving into.

It was November and the beginning of the trapping season. Frank left his house the morning of Thursday, November 10. He would be packing in by way of his brother-in-law Walter Stull's place, using the new Fire Patrol trail that ran five river miles to the Elkhorn Ranch, five miles shorter than the traditional pack route by Trail Butte. The fire patrol had a couple fire lookouts and telephone lines connecting them, and one of the telephone lines went through Stull's, coming up the river and on through the Elkhorn Ranch. The lines were almost never maintained during the winter months. Once Frank got to Elkhorn, he would use it as a staging area while he moved on up to the Bert Gould cabin, about five miles further on. From there he would establish his trapline as far as Gould's Lake, about seven miles further, on Elk Creek.

Cle Wilkinson, hunter and trapper.

When Frank left home Thursday morning he told his wife Margaret that he would be back on Sunday night. Sunday night came and went, then Monday night, then Tuesday. Margaret thought that was too much – he was three days overdue and she thought it would not be too soon to ask for help.

There was a remarkable response. Frank was a regular warden of the Coos Fire Patrol, and they were particularly concerned over his fate. All of the trappers who had knowledge of the upper West Fork responded. Cle Wilkinson

---

[7] Thanks to Keith Browning and Melissa Green for the family information.

[8] See *The Goulds of Elkhorn* by Aileen Barker Rickard (1982) for the full, fascinating story.

and others went from Glenn Creek into Gould's Lake to search downstream on Elk Creek. Others – including his relatives and in-laws went into the Elkhorn from the West Fork to search upstream. The CCC boys who were building roads into the Elliott were called into help. It was a major search, by any standards.

On the first day, the fire patrol hastily placed the telephone line to the Elkhorn back into service. These lines were a simple "ground return" system of a single galvanized wire strung from one tree to the next through porcelain insulators.[9] For the eighteen miles between the Elkhorn Ranch and Allegany Central those insulators were hung in thousands of trees, and then thousands more between Allegany and Fire Patrol headquarters at Bunker Hill. With the first storms of each season, the telephones were out of commission from falling limbs and trees, and they were usually not repaired until the beginning of the next fire season. On the first day of the search for Frank Bremer, the lines were quickly repaired to aid in communication for the search.

Even the best communications with headquarters will not find someone lost in the woods, however. There needs to be someone on the ground who knows what he is doing and this search party had one such member who stood above the rest. Fifty-seven year-old Clarence Gould, raised from infancy right on the Elkhorn, had intimate knowledge of all the trails, and was reputed to be among the best trackers in the county.

By mid-day Wednesday Clarence was on the search and began tracking Frank's trail out of the Elkhorn. He only got about a half mile when the tracks, which most woodsmen may have missed, left the trail and headed downhill. Frank's body was found below a rock bluff at the edge of the river. He was 38 years old.

Almost as soon as the body was found a telephone call was placed to Fire Patrol Headquarters in Marshfield with the news. Dr. Ennis Keizer, the county coroner, gave instructions to remove the body from where it was found and bring it to town. They carried it to Elkhorn Ranch for the night, and the CCC boys carried it out the next day, back down the new fire patrol trail to the Stull place, home of Frank's sister Caroline. From there an ambulance was waiting to take it to town and the Thuerwachter Funeral Home in Marshfield.

This was Thursday, November 17, and the story was in the *Coos Bay Times* that afternoon. Even more surprising, the weekly *Marshfield Sun* came off the press on Thursdays, and editor Jesse Allen Luse managed to get the story in the paper, almost scooping the big daily! And the *Marshfield Sun* was still using the excruciatingly slow hand-set type, the last newspaper in the State to do so. Frank Bremer "sets his last trap," was Jesse's epitaph for Frank, whose funeral was held that Saturday at the Allegany School.

[9] Jerry Phillips, op cit, pp 88-93.

Of course, the newspaper accounts were only the begining of the stories that would circulate about the event. "Frank Bremer is dead after fall on Upper River," is the way the *Coos Bay Times* put it. "Bremer is believed to have fallen to his death last Friday night while on a 13-mile hike between the west fork of the Millicoma and Elk Creek, setting traps."[10] There was no reported indication that there were tracks from anyone other that those of the deceased. Cle Wilkinson's daughter, Alice Allen, indicated that there was a general opinion that he had a stroke or heart attack, a seizure or aneurism, something of that sort.[11] Frank was subject to nose bleeds, Alice said, and nosebleeds could be indicative of high blood pressure.

The most persistent story is of the conspiracy type. Based on the assumption that, as traplines were not legally sanctioned entities, it might be assumed that they were therefore lawless. If anybody who was big enough could establish a trapline along one of these creeks, anybody who was bigger or more ruthless could take it away from him. Trapline robbers were not unheard of, closer to town, but out in these woods a generation of trappers had been respectful of each other's prerogatives. The gentleman's agreement was the law that governed their relationships. Trapline robbing was unheard of among them.

[10] *Coos Bay Times*, Nov. 17, 1938.

[11] Youst, op. cit.

And the theory of trapline robbers begs the question, what would they have been robbing? The traps hadn't been set. Frank was only a half mile from the Elkhorn, many miles from where the trapline would be established. And if there had been anyone else anywhere near, Clarence Gould would almost certainly have detected it. Jesse Allen Luse of the *Marshfield Sun* concluded, "There is no way of determining the cause of death, other than unforseen natural causes."[12] But such a conclusion does not make a very exciting story, and so the conspiracy theory persists!

Dr. Keizer did, however, perform an autopsy and he found multiple contusions on the body and legs. He concluded that the cause of death was pulmonary embolism from a leg bruise. The bruise, Dr. Keiser reported, had resulted from a fall after which Frank had apparently walked until the pulmonary embolism caused his death. The last question on the Death Certificate asked if it was in any way related to occupation of deceased. Dr. Keiser answered, "Yes. Trapper setting out traps on trap-line."[13]

No one moved their traps into the upper West Fork and Elk Creek for several years. At last, Cle Wilkinson quit his trapline on Matson Creek and moved into Elk Creek, bringing along his partner Baldy Crane. They planned to build a

[12] Marshfield Sun, Nov. 17, 1938, p. 3.

[13] Oregon State Board of Health Certificate of Death, Frank Bremer, Coos County, Nov. 12, 1938.

cabin and use it as a base for picking sword fern, which they could do almost year-round, and much more profitably than trapping furs. The fur trade as it had traditionally been known had just about come to an end, replaced by the "fur farm." Now it would be ferns rather than furs that were the economic mainstay of "minor forest products" to come out of those woods.

Margaret never remarried. She taught school, serving as principal at Allegany and later as a teacher in the Coos Bay schools until she retired, highly repected and well known.

The property remained in probate for several years, but in 1946 Margaret gained title to the entire 160 acres of William Bremer's original 1892 homestead. Most of it remains in the family to the present day, through her nephew Tom Huppi and neice Melissa Huppi Green, who live there.

Frank and Margaret Bremer's house on the West Fork c. 1940. Bessie Shivers, Beth Wilkinson Browning, Fred Huppi, Rose Browning Huppi. Photo courtesy Melissa Green.

Map V

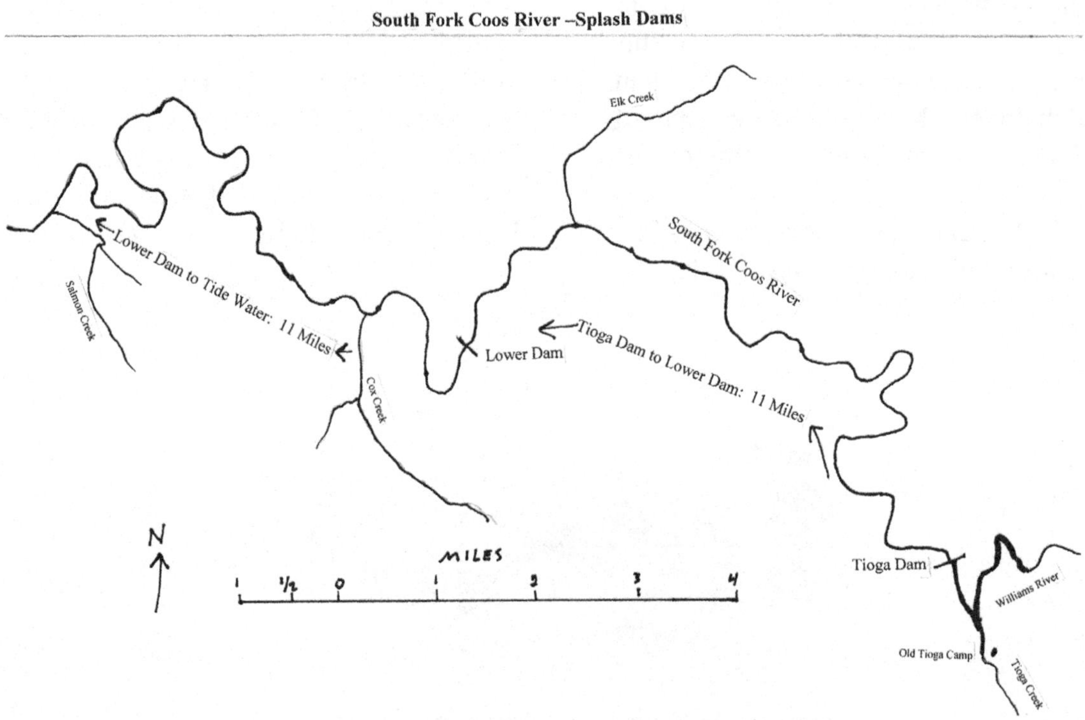
South Fork Coos River –Splash Dams
Elk Creek
South Fork Coos River
Lower Dam to Tide Water: 11 Miles
Salmon Creek
Cox Creek
Lower Dam
Tioga Dam to Lower Dam: 11 Miles
Tioga Dam
Williams River
Old Tioga Camp
Tioga Creek
N
MILES
1 1/2 0 1 2 3 4

Chapter 3

Splash Dams

Lost in the splash of log drives, 1925, 1945, and 1951

When I was a boy during the 1940's, we lived along the Millicoma River where they were still using splash dams to sluice the logs out of the woods to tidewater, where they were rafted and towed to the sawmills on Coos Bay. The Millicoma Boom Company had a franchise from the State to operate four splash dams on the river. The first one (as you go upstream) was about four miles above our house. On a regular basis we would hear the rumble of the larger old growth logs bumping along the bedrock for perhaps ten minutes before the splash arrived at our place. That gave us time to get out of the house and down to the road where we could watch the three or four thousand logs of a typical log drive pass us by.[1] It was very impressive, no matter how many times you watched it. Our parents continually impressed upon us the dangers, which were quite obvious even to a kid!

Before the hundreds of miles of logging roads were built into these woods, most of the logs came out by way of the river, that ancient highway. Logs float, and rivers run downhill, and floating logs down rivers to the mill was traditional around the world for several hundred years before loggers ever came to Coos County. But once they got here, they found the right geography for that kind of logging and they kept it up until 1957, probably the last place in the United States to use splash dams to drive logs.[2]

It was considered dangerous, and there are many stories of close calls. During the 1940's my dad, George Youst, had his sawmill on upper Glenn Creek, a tributary of the Millicoma River, and he built a dam in the creek for a mill pond and another dam, a splash dam, further

---

[1] See "Millacoma (sic) Boom Company" in *Coos Bay Times New Era edition*, 1937, p. 12, for a good summary of the splash dam operations on the Millicoma River. See also Dow Beckham, *Swift Flows the River*, pp. 73-7 for a more complete history.

[2] Dow Beckham reminds us in *Swift Flows the River*, p. 144, that Maine outlawed river drives in 1971, and that same year Potlatch Lumber Company made its last drive in Idaho. I do not know whether those drives used splash dams or if they used only the natural freshet. I suspect the latter.

up the creek to drive the logs to the mill. One evening he came home wet and unhappy, with an abrasion along his cheek. He had started to open the dam and was caught by a plank and flipped into the water. He went through the twelve-inch gap that was left by the plank and he miraculously emerged a couple hundred feet below the dam, wet and shaken but essentially unhurt.[3] Dow Beckam tells of several close calls in his book on log drives in Oregon and there are no doubt many that have not come down to us. But the theme of this book is search and rescue, and there are three incidents that I am aware of in which the search was on for persons assumed to be drowned or otherwise killed during a splash. One was on the Middle Fork of the Coquille River, and the other two were on the South Fork of Coos River.

Lockhart Dam, East Fork Millicoma River, 1952. This splash dam was 4 miles upstream from the Youst house.

Logs coming down in a splash, East Fork, Millicoma River, a few hundred yards upstream from the Youst house, 1940's.

The first was in June 1925, when a Hollywood movie entitled *The Ancient Highway* was being filmed on the Middle Fork of the Coquille River. A stunt man named Renald D. Jones, doubling for the star Tim Holt, was in a canoe below the Sugarloaf Dam. The scene called for him to be paddling ahead of a rush of water and logs. When the dam was opened, the wall of water immediately capsized the canoe and Jones was drowned in sight of hundreds of spectators. Filming stopped, the search was on, and his body was found the next day.[4] The film was released later that year to a *New York Times* review that summed it up: "Those who like logrolling and dynamiting thrills will find this picture leaves nothing to the

[3] My oral history interview with my dad in 1967 was published as Chapter 24, *Above The Falls, 2nd Edition* (2003), pp. 198-214. His account of going through the splash dam is on page 203-5.

[4] Curt Beckham gives his remembrance of the event in his pamphlet, *Tall Timber Tales*, Myrtle Point, OR, 1989, p. 7.

imagination."[5] There was no mention in either the reviews or the film's publicity about the death of the stunt man, and no print of the film is known to exist. Renald Jones died unheralded for the transient thrills of a movie-going public.

The next death associated with a splash in Coos County was on June 7, 1945. Glenn Thornton was killed about seven miles above the head of tide on the South Fork while helping to break a log jam. He was below the jam, hooking a cable to what he believed to be the key log when the entire jam began to move. Dow Beckham was river boss and almost lost his life in the same incident. He said, ". . . everything suddenly became hushed and quiet except for the muffled bumping of the logs and the squeaking produced by the logs rubbing against each other."[6] Glenn Thornton fell into the logs and was killed.

The search for his body was a very emotional undertaking by all concerned. Glenn was stepson of Oscar Lundberg, the road boss, and had been recently discharged from the Army after going through two years of combat in Europe. Germany had surrendered in May, and many of the veterans of the European part of the war were home. Glenn had been only two weeks on the job, but the war was still going on in the Pacific and affection for our veterans and active servicemen had never been stronger. To lose his life in a logging accident after going through the recently completed war in Europe had the effect of tragedy.[7]

The search lasted ten days. Dow Beckham told of fixing lights powered by 6-volt batteries to the ends of pike poles for searching the bottoms of the deeper pools at night. He said, "One fellow had cut a piece out of an old hot water tank to use as a diving helmet and attempted to explore the holes. He had no luck, though he walked around on the bottom for several minutes." At last it was decided to open the dams again to see if the body would be flushed out that way. It worked. The body floated up out of a hole that was about a mile below the area where most of the search had been concentrated. That was on June 17, just ten days after the accident.[8]

---

[5] *New York Times* movie review, Nov. 11, 1925.

[6] Dow Beckham, *Swift Flows the River*, Arago Books, 1990, p. 3.

[7] Glenn Thornton was a son of Signa Lundberg from a previous marriage. With her husband Oscar Lundberg there was a son, Albert, and a daughter, Maxine. They lived near us, and Albert and I were the same age, went to school together, and were close buddies when we were kids. Maxine was the maid of honor at my sister's wedding, which took place June 1, just one week before her stepbrother Glenn was killed.

[8] See Beckham, *Swift Flows the River*, pp. 1-5, and Charlotte Mahaffy, *Coos River Echoes*, p. 112, for accounts of the Glenn Thornton incident. The local newspaper concentrated on news from the recently completed war in Europe and the major buildup

The case of Frank Serus in November 1951, is the only other drowning in Coos County in connection with a log splash that I know of. He was a fern picker, and a stranger to the area.

Callison's Evergreens at 999 N. Front Street in Coos Bay always needed experienced fern pickers. The woods of Coos County had tens of thousands of acres of sword fern growing in the solid shade of the Douglas fir canopy, and those fern were much in demand by the international floral market. Placed in cold storage, they were transported by rail to the east coast and by ship to Europe. But not many who tried harvesting them for a living were successful because it was hard work and the market demanded quality and consistency. Bringing your first truckload of fern to Callisons only to have half of them culled for defects you hadn't noticed was reason enough to give up and try another profession. But for those who were good at it, it provided a living almost year-round.

And so it was on Wednesday, October 31, 1951, that Frank Serus, a 37-year-old fern picker, set up his camp three miles above the head of tide on the South Fork of Coos River. He had arrived from Crescent City, California, that day and had made arrangements for Phoenix O'Toole, manager at Callisons, to buy all his fern. Phoenix recognized a pro when he saw one. Serus was wearing the signature "forest green cruiser coat," the Filson product that in those days indicated he was among the coastal forest elite.[9] He set up his camp and the next day he picked 125 bunches. Then he disappeared.[10]

A "bunch" of fern consists of 52 individual fern fronds and in 1951 was valued at about 15 cents. If all the fern were good, his first day's harvest of 125 bunches would net Serus about $18.75. That is almost double what a man setting chokers at the Irwin-Lyons logging camp further upstream at Tioga would have made in 1951. I know, because in January 1951, I set chokers for Irwin-Lyons at their Tioga camp and if memory serves me correctly, I was getting $1.68 per hour and they took out $3.50 per day

---

of the war in the Pacific. There was almost no local news, and if there was any reference to the death of Glenn Thornton during the log drive, I could not find it.

[9] The green wool Filson cruiser coat was similar in style to the U. S. Army officer's "greens," patented by Filson of Seattle in 1914. It was jokingly called the "Coos Bay Tuxedo" and was considered appropriate dress in Coos County for even the most formal events. After the patent expired in 1927, it was widely copied by other clothing manufacturers. I bought my first cruiser coat for $25 in 1951 from the Oregon Woolen Mills store in Coos Bay. It was not an original Filson, it was a copy, and though it is badly worn out, I still have it (and I have an original Filson, which currently sells new for $279.50).

[10] I found most of the details concerning Frank Serus in two articles in the *Coos Bay Times*, "Fern Picker Said Missing," Nov. 8, 1951, and "Hunt for Fern Picker Ended," Nov. 12, 1951.

for board and room at the camp.[11] Frank Serus was quite a small man, only about 135 pounds, which may have placed him at a disadvantage as a logger. Picking fern was something he could do very well, and he was his own boss, working on his own time.

sword fern frond
52 fronds = 1 bunch

The sword fern was only one of the "minor forest products" that Callison's handled. There were other decorative floral evergreens: black huckleberry, salal, and cedar boughs, among others. And there were crude drugs, plants with medicinal value that were seasonally harvested in 1951: cascara bark (laxative), foxglove (digitalis), Oregon grape (antiseptic), and Douglas fir pitch. Those minor forest products were a larger industry in the Northwest than anyone realized at the time, running to tens of millions of dollars annually, mostly in cash.[12]

Typical patch of sword fern under an evergreen canopy.

Some of the men and women who went into the woods to harvest those products were pros at it and did nothing else. Others did it only to supplement income from other sources. Frank Serus was a pro, as evidenced by his having picked 125 bunches the first day – a very good harvest.[13] He was camped across the river from another fern picker, Herman Montgomery. They made acquaintance with each other and went on about their business.

Elk hunting season had opened, and as luck would have it Frank Serus ran across a herd of elk on Thursday, during that first day of work. He was apparently not interested in hunting but the next

[11] \$1.68 X 8 hrs/day = \$13.44. Subtract the \$3.50 for room and board and you get \$9.94 before taxes – about half what the fern picker made with his 125 bunches of fern at 15 cents per bunch.

[12] There is limited literature on minor forest products during the 1950's. See for example, *Secret Treasures in the Forests*, Oregon State Board of Forestry, 1951; *Harvesting Minor Forest Products in the Pacific Northwest*, by Thomas C. Adams, U. S. Department of Agriculture Pacific Northwest Forest and Range Experiment Station, November, 1960.

[13] In my October 18, 2003 oral history interview with Wilma Lund, a professional fern picker, I recorded her as saying that her daily goal was 100 bunches, a goal she only occasionally met.

afternoon, on Friday, November 2, he rowed Herman Montgomery's boat across the river and left a note telling Montgomery where he had seen the elk. Montgomery found the note, but the boat was missing and Serus's camp was unoccupied. His truck, camping equipment, and the 125 bunches of fern were undisturbed. Some time later the boat was found about one-half mile downstream with the oars in the rowing position. Still later, Serus' green cruiser coat covered with sand was found at the river's edge, near the camp.

It was a week later, November 8, that the State Police were notified and a search party of six men began working the river. As soon as the Irwin-Lyons river crew found out about it, river boss Dow Beckham had his crew working full time for several days, searching waist deep in the cold water. There were as many as twenty men at a time on the search, with an estimated total of 50 men who were active at one time or another. When Phoenix O'Toole at Callison's was notified that Serus was missing he immediatley posted a $100 reward for information about Serus, or finding the body.[14] But after four days of search (and eleven days afer he had turned up missing), the search was called off for Frank Serus. His truck and camping equipment were removed to Phoenix O'Toole's home in North Bend where it remained while Phoenix tried to locate next of kin. His body has never been found.

There was good reason why Dow Beckham and his river crew would want to be engaged in the search. As it happened, they had released the impounded waters of the two logging splash dams of the South Fork on Friday, November 2, the same day that Serus turned up missing. What probably happened seemed so obvious that it remained unstated. The boat was probably struck broadside by the initial bore of the splash – a two or three foot head of water, completely unexpected by Serus rowing the boat across the river. He could easily have been thrown off balance and fallen into the river when the bore of the splash hit the boat. Splash dam log drives had been used on the South Fork of Coos River since 1941, and on the East Fork of the Millicoma River since 1884. As dangerous as it apparently was, this is the only case that I know of where someone downstream, and unconnected with the splash, was caught in a splash and drowned. Frank Serus was newly arrived from Crescent City and he may not have even known that the South Fork was subject to

[14] Phoenix O'Toole apparently maintained very close relations with the regular fern pickers. I interviewed Wilma Lund for the Coos County Oral History Project in 2003. She was a fern picker of many years experience and spoke highly of Phoenix's loyalty to her and other pickers. At one time a pair of thieves stole a load of her fern (which she recovered), and she said that Phoenix told her that he would have recognized her ferns and the theives would not have gotten away with it. Interview Lionel Youst with Wilma Lund, Oct. 18, 2003.

frequent log drives, sluicing the logs out of the Tioga country. The two forks of the Coos River may have been the only places in the United States where that primitive means of transporting logs was still in use at that time.[15]

To have been crossing the South Fork at the very moment that the initial bore of the splash arrived at that very spot, was unlucky, really unlucky, to say the least. If there is one element that seems to drive the events in all of these search and rescue stories, it is luck. With it you survive, without it you don't. At that time on that day, Frank Serus didn't have it, nor did the stunt man Renald Jones on his day, nor did Glenn Thornton on his.

Tioga Dam, 22 miles above the head of tide, South Fork, Coos River. Notice logs in the impounded water behind the dam. The dam drove millions of logs during the years 1944 to 1957. Photo courtesy CCHMM, 988-NA13.

[15] The history of splash dams on the South Fork of Coos River is superbly narrated by the river boss himself, Dow Beckham, in his book, *Swift Flows the River: Log Driving in Oregon*, Arago Books, 1990. In it he describes the "Lower Dam," built in 1940 and 1941, located eleven miles above the head of tide, and the "Tioga Dam," built in 1943, located 22 miles above the head of tide – just below the confluence of Williams River and Tioga Creek. When those dams were removed from service in 1957, it was the end of splash dam logging in the United States.

Map IV
Lake Creek and Crash Site

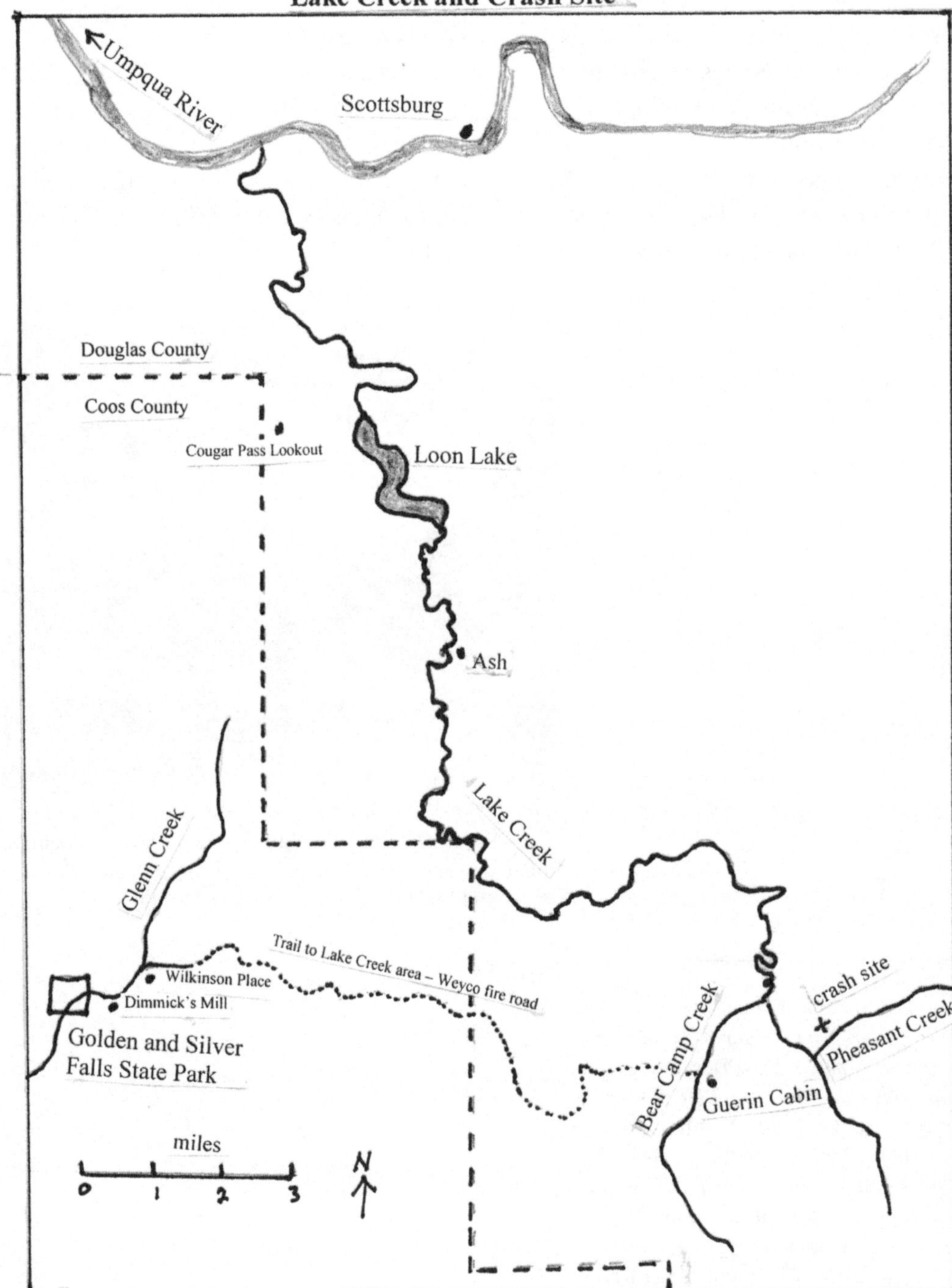

Chapter 4

An Army C-46D. Down With 12 Men Aboard

This chapter was published in its original form in *Above the Falls,* pp. 289-307 (1992) and revised for *Above the Falls*, 2nd Ed. pp. 232-251, (2003). I have here revised and reformatted it from the 2nd Ed.

It was the last week of November 1945, and off the Oregon coast another storm was approaching. This would be the first big one of the season, and it would be a doozy! It would last six days and its gale winds would drive eighteen inches of rain into the woods above the falls. Because of it, the general public for the first time would become conscious of how large, how wild, and how remote these woods really were.

The great Douglas fir forests of the Coast Range have succeeded themselves for perhaps sixty generation – 10,000 years or more. They are sustained by an annual 60 to 100 inches of rain. Most of this rain comes from the southwest, off the Pacific Ocean from which it arrives by a succession of storm fronts during the six months between October and April. It rains during the other months as well, but the big storms and the large rainfall occur during that time. The result is a temperate rain forest with a prodigious vegetation of mosses, ferns, fungi, shrubs, and trees of amazing variety, linked together in an ecology no one has ever yet figured out.

Before the days of weather satellites, the storms often came by surprise. The barometer would fall and sometimes within a few hours a hurricane-type storm would be on us. Nowadays, the nightly TV keeps us informed of the progress of these fronts across the Pacific. Not so in 1945.

World War II had ended three months earlier, when the Army Air Force dropped the atomic bombs on Hiroshima and Nagasaki. The business of reorganizing for the post-war era was just starting. Tens of thousands of men were being discharged and equipment was being transferred for sale, scrap, or storage. It was an unsettling experience, for peace to break out after four years of war.

At McChord Field, near Tacoma, Washington, there were six L-5 aircraft that someone in the Air Force had decided should be moved to Eagle Field, California. The L-5's were single-engine liaison planes built by Stinson-Vultee during the war years. They had a 185-hp O-435 Lycoming engine and they were used for ambulance, cargo, aerial photography, and general transportation. They were good utility planes, and the Air Force did not declare them surplus at the end of the war. They were still using some of them in the 1960's.

To fly those six L-5's to California, someone else in the Air

Force had decided to use six pilots who were at Sedalia, Missouri, and Denver, Colorado. On Saturday morning, November 24, 1945, a Curtis C-46D "Commando" departed Sedalia on the first leg of a mission to deliver those pilots to McChord. The C-46 had two 2000-hp Pratt and Whitney R-2800 engines and a 48,000 pound gross weight. They were good cargo aircraft and the Air Force had received 1,410 of them during 1943 and 1944. It was not a small plane by any means. It was, in fact, the largest twin-engine aircraft in production during its time. It had an 1800- mile range with 8.5-hour endurance with full fuel load.

During the very early morning hours of that fateful Saturday the crew chief, Pfc Walter Smoyer, ran his pre-flight check list. The aircraft could not have been more than two years old and was no doubt in excellent condition. Smoyer, like many good crew chiefs before and after him, would have taken great pride in the excellent condition of his "bird." He would have cranked up the two Pratt and Whitney "Double Wasp" engines and they no doubt "purred like a tom cat!" These were fine engines, well cared for. All they needed now was fuel.

At the terminal there were two men who were looking for a ride to the West Coast. Corporal Everette Bailey was on leave and looking for a ride home. Major Frank Gaunt had been discharged from the Army and was on terminal leave, also on the way home. There was plenty of room. The C-46 could carry twenty-eight passengers and a crew of four.

Probably an hour or so after Pfc Smoyer arrived, the other three crew members would have shown up. They had already filed their flight plan, which called for them to fly to Denver to refuel, then to Ogden where they would remain overnight. When the crew showed up, each member had his own pre-flight check list to run. The pilot, Captain Hugh B. McMillan, the co-pilot, Captain Harlow B. Marsh, and the radio operator, Sergeant Robert T. Neal, each had their individual responsibilities. Everything checked OK. Things were looking good for this mission.

In those days, as soon as the passengers were manifested, they were fitted and issued parachutes. Nowadays this would only be done for flights on combat aircraft, but back then the Army thought the passengers needed a way out. While the crew was running its pre-flight inspection, the passengers would wait in the terminal with baggage and gear.

It was a routine flight, but a long one. At Denver they refueled and picked up four more passengers, then proceeded to Ogden where they stayed overnight. The next morning (Sunday, November 25), they flew over the Sierra Nevada Mountains at 16,000

C-46D Commando

feet, without oxygen or pressurization, to Fresno, California, and then on to Oakland. Smoyer felt "pretty sick" after about two hours at that altitude, but that evening he and Sergeant Neal went to San Francisco where they had a great time. Sergeant Neal called his fiancee in Los Angeles and told her they were leaving for the north in the morning. He had been to San Francisco once before and was able to show Smoyer around. They rode the cable cars and took a ferry ride across the bay. Smoyer said the whole evening only cost him $1.05, and he had a great time.

In the morning (November 26) the four crew members loaded their two hitchhikers, Corporal Bailey and Major Gaunt, and the six flight officers who were scheduled to fly the L-5's back from McChord: Robert T. Kenneth, Dave Reid, Floyd C. Waddill, Ralph Foster, Theodore Hartsong, and Second Lieutenant Jonathan D. Clark. At about eleven o'clock they took off for a routine flight to Portland.

Back up above the falls at Allegany it started to rain during the evening before, and this morning it was obvious that a good one was coming. Cle Wilkinson normally rose at daylight, and when he got up he could see what they were in for. He was batching these days. His brother-in-law, Alfred Leaton, was staying with him at the time. His mother, Laura, quite elderly now, was living with her daughter in town. Cle's daughters were both away at school. Pat was staying with her aunt and uncle in Coos Bay, and Alice was at college in Monmouth. Cle got up and cooked breakfast.

Vic Dimmick's mill on the Tyberg place was still running. George Youst's mill had been shut down for several months. Weyerhaeuser had started building a fire access road up Woodruff Creek from Cle's place, along the old trail to Lake Creek. Logs from the right-of-way had been sold to Dimmick and cut in his mill. The road had been punched through about eight miles and there was a small trailer house at the end of that road, where dynamite was stored. The Weyerhaeuser presence was not obtrusive, yet, but everyone knew that now the war was over, they would be building a big mill at Coos Bay. Big changes were due in the land above the falls.

Below, the old Tyberg place was now owned by Roy Grant. Vic Dimmick's sawmill was in the field with a dam in Glenn Creek, which backed up water for the mill pond. Roy Spires was the timber faller and high climber for Dimmick. It would be too windy for him to work today, so he would take care of things around camp and at the mill. Most things were going as usual that Monday morning in November 1945, but it was clear that we were already in the middle of a very big storm. Offshore, seas were extremely high. There was a *big* blow behind them.

At Oakland, California, there was no indication that such a storm was hitting the Northwest Coast. California was in sunshine. It was only about three hours flying time from Oakland to McChord, and so the C-46 probably took off with enough fuel for six hours endurance. As they flew north, they were east of the Coast Range and the worst of the storm did not affect them. But before they got to McChord, they knew they were in trouble, and by the time they arrived, no aircraft was allowed to land. This was probably a little before 2 p. m.

The control tower at McChord ordered them to Pendleton, according to first reports released by the Army. Pendleton was clear. Later, the surviving passengers said they had been ordered to Portland. Radio reception was bad, and the crew may have understood the tower to have said "Portland" when in fact it said "Pendleton." We'll never know for sure, but it was fatal.

Arriving over Portland thirty minutes later, they could not land there either. The control tower at Portland ordered the plane to return to Oakland, and the crew accepted the order and continued south. They had enough fuel left for two more hours. Oakland was almost three hours away, even had there been no headwind. But there was a headwind, and as it turned out there was only two hours of fuel left.

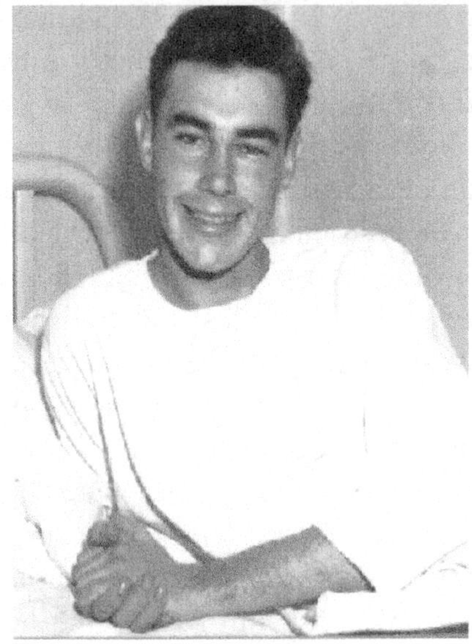

Flight Officer Ted Hatsog of Greenwood, South Carolina

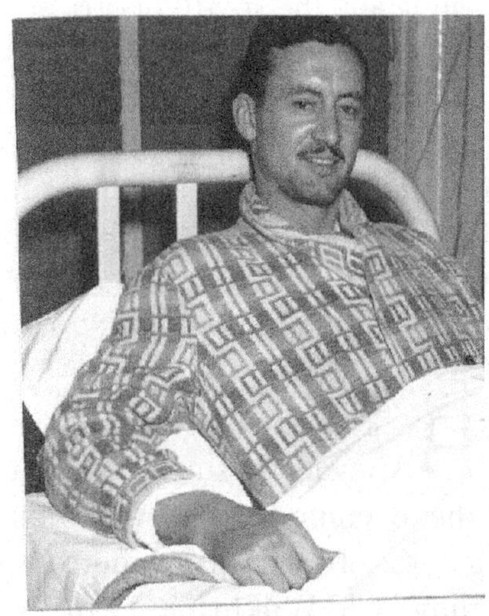

Major Frank Gaunt

Note: These eight pictures and the one on page 45 were taken by the Coos Bay Harbor photographer at the McAuley Hospital in Coos Bay.

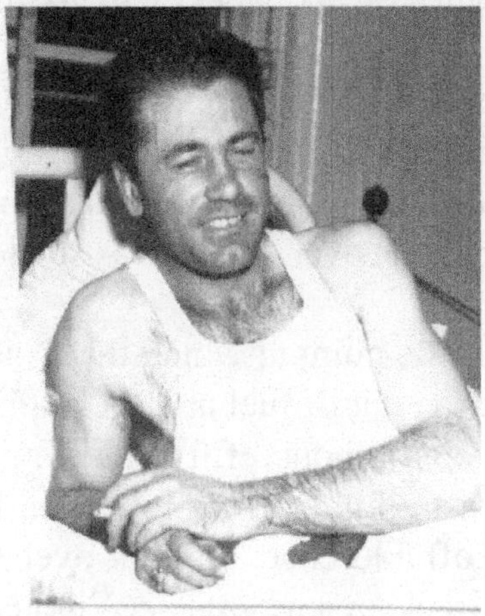
Flight officer David Reid of Altus, Oklahoma

Flight Office Floyd Waddill of Laplata, Missouri

Flight Officer Ralph Foster of Witchita, Kansas

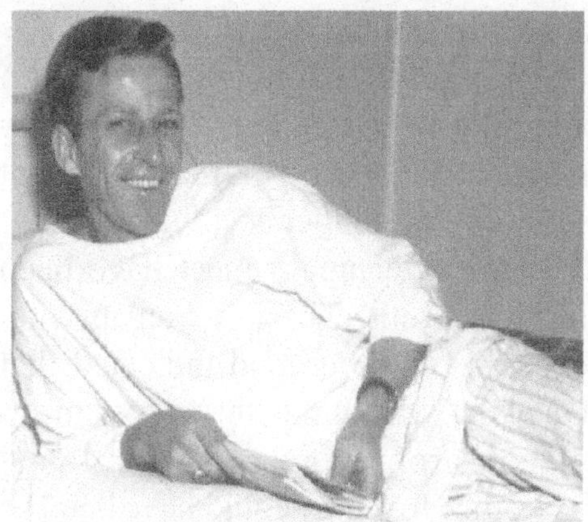
Corporal Charles Bailey of West Palm Beach, Florida

2/Lt Jonathan Clark of Rogerville, Tennessee

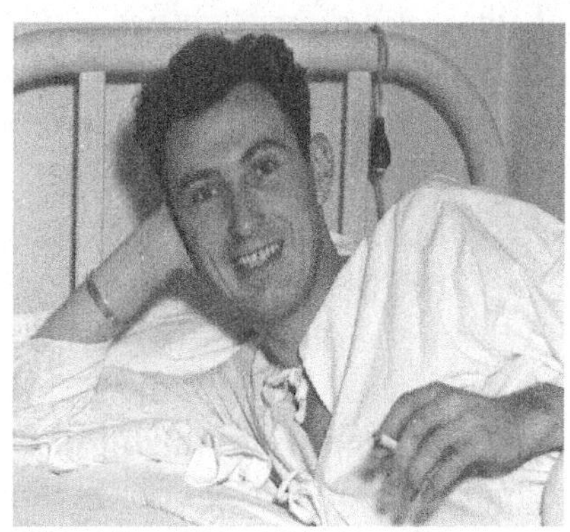
Flight Officer Richard Kennett of Chicago, Illinois

The storm had increased and promised to be perhaps the biggest of the decade. Slides had closed the roads from the valley to the coast. The Coos Bay - North Bend water system had washed out. Highway 101 was flooded between Coos Bay and Coquille. A barge had broken loose from its tug and was adrift offshore of the Umpqua River. No ships could cross the Coos Bay bar. Coos River was at flood stage, and the Enegren Ferry could not cross. Allegany residents who were in town were stranded there. Schools at North Bend and Coos Bay would be closed for the rest of the week.

Radio communications on board the aircraft failed. One can only imagine the attempts that Sergeant Neal, the radio operator, was making. Their radio compass must have been faulty as well. The last transmission received by the ground indicated that the plane was about thirty-five miles offshore from Florence. This was at 3:15 pm. How they could have been so far off course remains a mystery, but it seems they were flying right into the storm all the way from Portland. I'm told that with the radio navigation aids of that time, it would have been quite easy to get on the wrong side of the beacon and not be able to find it again. That may have been what happened.

By that time there was only fifteen minutes fuel left. The two Pratt and Whitney dual-wasp R-2800 engines continued to purr along, but they couldn't keep it up much longer. A heading correction was made, and they continued southeasterly. Passengers and crew had their parachutes on and were ready to jump when ordered. Major Gaunt, on terminal leave after four years of war, had already bailed out of two disabled aircraft, and this would be his third. The war was over and he was so close to home.

The point at which the C-46D finally ran out of fuel and the engines quit was about fifty-five miles southeast of their location over the ocean off Florence. It was over the Lake Creek area where Cle Wilkinson had his trapping cabin and where the Guerin's and other Lake Creekers had homesteaded thirty-five years earlier. It had been a remote wilderness then, and it was still a remote wilderness.

This was just a few minutes before 4 p. m. The engines quit and the aircraft descended below the clouds. At a little over 1000 feet there was a brief clearing in the weather. Passengers and crew saw a small meadow below them, in the midst of a forest which stretched to the horizon. That meadow must have been the site of Guerin's cabin, above Bear Camp Creek. Captain McMillan, the pilot, turned toward the meadow and gave the order to bail out. Major Gaunt acted as "jump master," and was the ninth man out. The tenth and last man out was the crew chief, Pfc Walter Smoyer. Smoyer said that Sgt Neal, the radio operator, was very scared, and asked if he should count to four before pulling the rip chord. "I told him to pull it right away as we were very low."

Smoyer said, "I went out last and my chute opened and I started to sail on to the ground. My chute caught on the top of a giant fir tree that must have been 250 feet high. I looked at my watch and I landed in the tree at 4 p. m. I heard the plane crash and then everything went silent. There I was all alone just hanging in my chute and scared to death that my chute would blow off the tree. The wind blew me into the trunk of the tree. I tied myself to the trunk of the tree that night. I don't think I slept but I did doze off a little."

Major Gaunt's chute opened just before he hit the ground, and he heard the crash of the aircraft at the same time. Sunset that day was at 4:27 p. m. When the sun goes down in those woods, it is dark. It is especially dark during a storm. The flyers had less than thirty minutes of light. It was fourteen miles by trail, if they could find the trail, to the closest human being - Cle Wilkinson. The flyers didn't even know what state they were in. The Coast Guard and Air Force thought the plane went down at sea. The worst storm in memory was closing in on them.

At Brookings there was 6.28 inches of rain that night. At North Bend there was 4.07 inches. The hills above the falls typically receive almost double the rainfall of the coast, so the full extent of this downpour can only be imagined. The creeks and rivers were raging torrents.

If I were capable of writing a full account of the amazing rescue of these flyers, it would fill a book. I would give the account of how, twenty-four hours later, Flight Officers Waddill and Hartsog stumbled onto the Cle Wilkinson ranch, telling Cle that David Reid was stuck in the top of an old-growth fir tree fourteen miles to the east. I would tell how Cle went down to the Dimmick sawmill at the old Tyberg place and got Roy Spires, the highclimber. Hartsog then led Cle, Cle's brother-in-law Alfred Leaton, and Roy, back through the woods, in the dark and storm, to Major Gaunt and Flight Officer Foster, who were waiting at the Weyerhaeuser "dynamite" trailer. I would then tell how Major Gaunt led the rescue party another three miles in the dark to the exact tree that Reid was in.[1]

Reid had been hanging 150 feet up in the crown of a 200-year-old Douglas fir for thirty-six hours. During that time the rain and wind had hardly let up for a moment. At this point it would be necessary for me to describe one of the most astonishing rescue feats on record. Still in the dark and the storm, Roy Spires, a fifty-year-old logger, put on his belt and spurs and climbed the 150 feet to the first limbs of the tree. Once there, he sat on a limb, took off his climbing gear and put it on Captain Reid, who by then was paralyzed with cold and exposure. I would then describe how, miraculously, Reid had somehow made it safely to the ground. Spires pulled his belt and spurs back up with

---

[1] *Oregonian*, Nov. 30, 1945, p. 1, has the story and pictures. Reid landed in the top of the tree at 4 p.m. Monday and was rescued at 4 a.m. Wednesday, 36 hours.

his pass rope, donned the equipment, and himself climbed down, the seven of them returning the fourteen miles through the woods to Cle's house – still in the rain.

Waddill had, I believe, suffered a mild concussion when he landed. He was taken to town by someone from the Dimmick mill. When he got to the Enegren Ferry, which crossed the lower Coos River, they couldn't get across because of the high water. Telephone lines were still working, however, and they contacted the John Milton ranch. Milton relayed the information to town. A tug boat was pressed into service to help the ferry cross the river.

This was the first anyone knew that the plane hadn't gone down in the ocean thirty-five miles west of Florence. It was now reported on the radio that the plane crash had been somewhere "about fifty miles east of Coos Bay," and that there were survivors.

My brother-in-law, Jerry Baughman, heard about it on the radio. He decided to drive out to the Enegren Ferry to see what was going on. There were several Army officers and an ambulance waiting there. Word had it that there were three more flyers at Cle Wilkinson's place. Did anyone know where that was? Jerry said he did, so they asked him to guide the ambulance. It was after dark, now fortyt-eight hours since the crash. Jerry told the driver to be carful as they climbed, through the darkness, over the cliff road of Golden and Silver Falls. The rain was turning to sleet that almost stuck to the windshield as they crossed the Silver Creek bridge.

At Cle's place there were a number of loggers ready to go into the Lake Creek country to find any other survivors. No one knew whether there were any. Among the loggers there were several who knew quite a bit about the area, even though it would be several years before any logging would take place there.

At daylight, now sixty hours since the crash, Jerry Baughman returned in the ambulance with the three flyers who were at Cle's – Foster, Reid, and Major Gaunt. When Jerry told them that the Coast Guard had reported them lost at sea, the flyers were shocked. It was the first they knew that they had been thirty-five miles over the ocean only fifteen minutes before they ran out of fuel. After they got to the ferry, Jerry returned to join the search party, which by this time was growing. Fifty Army and Air Force personnel from Camp Adair and Portland Air Force Base were on the way.

State Game Warden Harold Stromquist was among the searchers. He was in charge of a contingent of men from the Coast Guard. With the possible exception of Hitler and Tojo, Stromquist was probably the most hated man any of the loggers and ranchers knew of. He was hated not so much because he arrested those he caught violating game laws. Any game warden would do that, and all poachers understood it. Stromquist had earned a special degree of contempt because of his practice of enticing children to talk about their home life and thus unwittingly incriminate their parents. I remember that he once he grilled me about our diet at home, with special emphasis on how often we ate venison.

Children in the Allegany area were taught never to use the words "deer meat," "venison," "elk meat," etc. We could take the Lord's name in vain, but we couldn't use words relating to game lest we inadvertently use the same word in front of Officer Stromquist. My parents didn't even hunt, but there was the danger that we might cause a neighbor to come under suspicion. Stromquist was feared and hated.

Jerry Baughman was in the group with Stromquist. He told me that the first night they were out, everyone was freezing cold and wet. The loggers had all brought extra socks along with them, but Officer Stromquist hadn't. Stromquist, with his wet and cold feet, looked longingly at those spare, dry socks the loggers had brought and said that he would "sure like to have a dry pair of socks." The loggers wouldn't give him any. He begged. The loggers teased him, asking if he would let them go if he caught them with illegal meat. He said, "Of course not. I can't do that." "Sorry," said the loggers. "No dry socks!" And Stomquist went through the search with wet, cold feet. Afterwards, Cle Wilkinson observed that Stromquist did a very good job on the search. He was knowledgeable of the area, and he knew what he was doing. Cle had as much reason as anyone to dislike Stromquist on principle, so praise from Cle carries special weight. Actually, everyone who was on that search acquitted themselves well. We can't fault any of them.

By this time the newspapers were picking up pieces of the story. Knowledge of the geography of this great forest was slight indeed. It was one vast *terra incognita*, as far as the outside world was concerned. Reporter Leroy Inman, who covered the story for the *Coos Bay Times*, gave the following description on November 29 (page 1):

> Whereabouts of the remaining passengers were not determined, but it is believed at least some of the men are alive and trying to reach civilization. It is believed they may have landed on the opposite side of a ridge and may be following the course of Bear Creek, which makes a wide circle, finally coming out to a road at Stolls Falls. The round-about journey would take a man an estimated two weeks, unless they can be contacted earlier.

It is hard to imagine a more distorted view of the lay of the land, but it is an excellent example of how little was known by anyone other than the few trappers and woodsmen who had actually been there. Stulls Falls are on the West Fork, which flows ultimately into Coos Bay. Bear Creek (Bear Camp Creek on the maps) runs into Lake Creek, which flows into Loon Lake, which in turn empties into the Umpqua River between Reedsport and Scottsburg.

Cle Wilkinson's old trapping cabin was on Bear Camp Creek, not far from its confluence with Lake

Creek. It was the swollen waters of Bear Camp Creek that Floyd Waddill had parachuted into, almost drowning before he got out. Waddill's experience was a close call, but so were the others. Three of the four remaining flyers who had successfully bailed out landed in the top of the old-growth Douglas fir. Their stories are as interesting as adventure stories can get. They are each different.

Flight Officer Richard T. Kennett landed in an open space about twenty feet square, between the trees. He turned his ankle, but was otherwise uninjured. Kennett was the only one of the group who had ever previously seen Douglas fir. Neither he nor any of the others could believe, however, that trees could be this large. Some of the men thought they had landed in the California redwoods.

Corporal Charles E. Bailey landed high in one of the trees, hanging in mid air, suspended by his shroud lines. It took him about twenty minutes to unhook his leg harness. Still hanging by his shoulder harness, he swung himself to the top of a smaller tree. Once in the smaller tree he unfastened his shoulder harness, leaving his chute in the first tree as he climbed down the second, limb by limb. When he reached the lowest limbs, still too high to jump, he was able to catch hold of the top of a tall sapling, on which he swung, unhurt, to the ground. He had been within earshot of Kennett during much of this time and after about twenty-five minutes - well after dark - they found each other.

Lt. Jonathan D. Clark, of Rogerville, Tennessee, must have been a cool character. He landed in the very top of a very large old-growth fir. Not knowing exactly what his next best move might be, he filled his pipe, lit it, and casually had a smoke in the rain and storm, 250 feet off the ground and fourteen miles from the nearest road. Finishing his smoke, he got out of his parachute harness and started climbing down through the crown of the tree.

Below the crown this was an unusual Douglas fir. As an even-aged Douglas fir forest matures, the trees tend to prune themselves. The limbs below the crown die, break off, and ultimately disappear. The tree trunk grows out around the wounds of the old limbs, and it is typically 100 feet or more of smooth tree trunk to the first limb. All signs of the knot are normally gone. Sometimes a tree might be standing a bit more to itself, and pruning would come later. In this case, pin knots - knots that were dead but still adhering to the trunk of the tree - would stick out from the trunk. The pin knots may be as short as three inches, and up to several feet long. Some of them would fall off if they were touched. No one who knew the Douglas fir forests would think a man could use the pin knots to climb to the ground.

The tree that Lt. Clark landed in had pin knots all the way to the ground. He threw down his parachute, shoes, socks, and camera and started down on the pin knots. An hour and a half later - well after dark - he was safely in the ground in the company of Kennett and Bailey. He was only able to find one of his shoes.

That first night the three men huddled next to a tree, covered with the parachute. The rain and storm increased, and after a few hours the three men began uncontrolled shivering, which didn't stop for six days. Their matches were wet and their cigarette lighters were dry. They could build no fire.

Pfc Walter A. Smoyer, the crew chief, had the hardest time of all the survivors. He landed in the crown of a tree evidently some distance from any of the others. Alone and in the dark and storm, he tied himself to the branches, where he swayed in the wind, soaked to the skin, until daylight. At that time he recognized that he was strictly on his own. He cut himself loose and used part of a shroud line as a safety rope, climbing down to the last limb, which was about fifty feet above the ground. His "safety rope" was too short to reach around the tree at that point, and it seemed his only chance was to let himself go. He cut the safety rope and went down, feet first, being knocked unconscious for awhile upon hitting the ground. It had taken him four hours to get down.

He wasn't too far from the raging waters of Lake Creek. Smoyer followed it through almost impenetrable brush as best he could. Freezing with wet and cold, he made about two or three miles per day. About six miles north of his starting place and still in the rain, he crawled out from under a log eighty-four hours later and he could go no further.

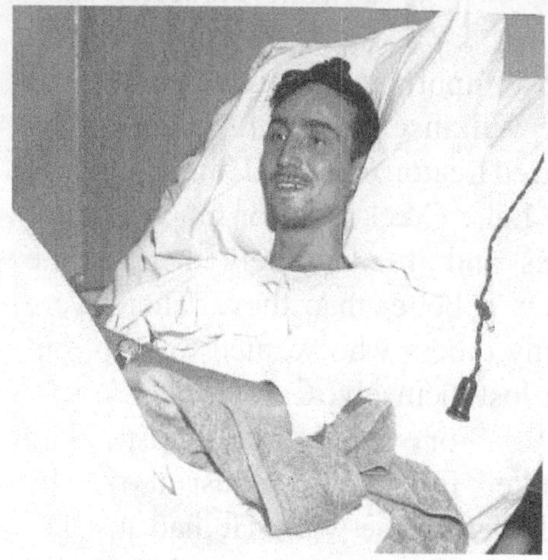

Pfc Walter Smoyer of Bethlehem, PA

Kennett, Clark, and Bailey had better luck. They stumbled onto an abandoned trapper's cabin Tuesday morning, where they found a can of carbide. Not many people would know what to do with carbide, but Lt. Clark – the Tennessean who smoked his pipe in the top of the tree – knew what to do with it. He punched a hole in the top and poured a little water in it. As the gas escaped, he lit it with the flint from his cigarette lighter, and they had a bit of fire. That night they rested a little better.

The next day, Wednesday, they came upon Cle Wilkinson's old trapping cabin near the junction of Bear Camp Creek and Lake Creek. In the cabin they found matches, three potatoes, a half gallon of meal, half a cup of bear tallow, and some syrup. They converted this into french fries and hotcakes, of which Bailey said later, "This was the best meal I'd ever eaten!"

Meanwhile, the search party was growing larger. Most important among the guides was Cle Wilkinson and his brother-in-law Alfred Leaton. They had run traplines in the Lake Creek area for much of their lives and it is unlikely that anyone knew it better than they. There were many others who wanted to help, but got lost themselves.

For example, Hank Gosney had worked for George Youst above the falls before the war. He had just got out of the Army and he thought he could do some good. He walked into the Lake Creek area but after being lost without food for two days, he and his partner finally stumbled onto Cle Wilkinson's place, cold and hungry. No one was home. They went inside the house and cooked all of Cle's bacon, left a little money, cleaned up their mess, and departed.

Cle's place was used as a kind of headquarters for the rescue operation. Radio contact from Coos Bay could not be made beyond the ranch, and so it was considered to be at the very edge of the known world. From the beginning, Cle was in the woods, advising the leaders of the search. The masses of people who later came up to be part of that search ended up taking over Cle's house.

On the third day a contingent of thirty-five personnel from Camp Adair and Portland Air Force Base arrived. The Army bought them each a new pair of caulk boots. The spikes on a new pair of caulks are razor sharp and are only worn in the woods, normally. The young Army men, however, were oblivious to custom and boldly walked all over Cle's house with their caulk shoes as though it were a barn. They did this over a period of at least four days. When Cle returned, his floors were ruined. The wood floors were punctured with thousands of holes and splintered beyond repair. The linoleum was cut to ribbons. His daughter Alice remembers that the Army paid him $150 in compensation for the use of his property.

On Thursday, November 29, Kennett, Bailey and Clark headed down Lake Creek. They assumed, correctly, that a stream had to come out someplace. They didn't know, however, that with Lake Creek at flood it would not be possible to follow it. The Lake Creek valley was an impenetrable jungle of matted brush. They went perhaps two miles the first day, then decided to return to the cabin and build a raft. They didn't know, but a raft trip would probably have ended in disaster at the forty-foot Lake Creek falls.

On Friday, about 9 a. m. while they were building the raft, they heard an airplane overhead. It was a Coast Guard PBY. The parachute was already laid out on logs so it could be seen from the air. The crew of the PBY spotted it, dropped walkie talkies and K-rations, and reported the position to the command post. The PBY then returned to North Bend and picked up sleeping bags, food, clothing, and heavy shoes.

The Army officers in charge insisted that what was really needed was toilet paper and water. Cle pointed out that there was no shortage of ferns and brush to wipe on, and it had rained eighteen inches since the beginning of the search. In Cle's judgement, if there was anything they did *not* need, it was toilet paper and water!

There had been notices on the radio saying that experienced woodsmen were still needed. As usual in these kinds of situations, there was a notable lack of communications. Dow Beckham and his crew of about eight men were coming home from work at Irwin and Lyons logging camp on the South Fork. Dow stopped at the Enegren Ferry and asked an Army officer if they still needed any help. The officer replied, "Gosh no. We don't need any more characters running around up there!"

Two more characters did show up, however. It's a good thing they did because they are the ones who found Pfc Smoyer and saved his life. Early Friday morning Elvin Hess and Archie Clawson, loggers from Allegany, went up to the Wilkinson ranch to see if they could help. They knew the country quite well. Elvin had fallen timber for George Youst at the Wilkinson place before and during the war, and Archie was raised on the East Fork. Both had packed into Lake Creek many times on fishing trips.

Jeeps on the Weyerhaeuser fire trail to Lake Creek.

Soldiers from Portland Air Base and Camp Adair, in the rain at the ferry,

Wheeled vehicles couldn't make it over the Weyerhaeuser fire access road up Woodruff Creek any more. It was a rutted river of mud from the Army jeeps and other four-wheel drive vehicles that had already attempted it. Elvin and Archie walked the eight miles to the "dynamite" trailer, which had become something of an advanced command post. Army Lt. DeMers was in charge. Harold Noah and a few other loggers were there.

About the time Hess and Clawson arrived, word came over the two-way radio that the Coast Guard PBY had located three survivors. They were said to be about "three miles away, on Pheasant Creek." The whole group headed out - no food, no bedding - expecting to be back in a few hours. After awhile the PBY radioed that it was not Pheasant Creek, but Lake Creek that the survivors were on. It also turned out that it wasn't three miles, but more like six miles, through the brush.

At that time, Hess and Clawson left Harold Noah with the rest of the group and went on ahead. They crawled on their hands and knees through the tangled brush of Lake Creek until, about a mile north of Bear Camp Creek, they saw the footprint of a man, along with a gum wrapper. They continued on, Hess firing his rifle every few minutes until about six miles north of Bear Camp Creek he found a track not more than twenty-four hours old. He yelled, fired his rifle, and heard a faint response. After searching in circles for a time, they came upon Smoyer, still sitting in the position he had assumed early that morning when he had crawled out from under the log. He couldn't move. This was at 4 p. m. Friday afternoon and ninety-six hours after the men had bailed out of the plane.

Hess and Clawson built a fire to warm Smoyer, but they had no food or warm clothing for him. Clawson stayed with him while Hess headed back upstream to the point at which the PBY was circling. He crossed Lake Creek on a fallen tree and came to Cle's old trapping cabin. Kennett, Bailey, and Clark were still there and, by coincidence, had now been joined by the trapper, Gard Sawyer.

Cle Wilkinson hadn't been trapping Lake Creek for several years now. He had pulled up his traps and moved them to Matson Creek in about 1936. Later, he moved over into the West Fork of the Millicoma after Frank Bremer died while running his trapline there in 1938 (see Chapter 2). It is amazing, but it took almost fifty square miles of wilderness to keep a trapper going. One trapper for each major drainage was about right. A few years after Cle pulled out of Lake Creek, Gard Sawyer, a descendant of an early settler on the Umqua, moved in. He came in and out by a trail from Elkton and as chance would have it, arrived at his cabin with a packsack of provisions just before Elvin Hess arrived.

Gard Sawyers cooked up a feed for the famished flyers, a feed they raved about later when interviewed by the press. The PBY had dropped sleeping bags, flying suits and K-rations at the cabin. Hess asked for a first aid kit, which the PBY dropped, but it landed in the top of a tree about two feet through. Hess chopped the tree down with the small ax at the cabin, but it hung up in another tree too large to chop down, and so the first aid kit could not be retrieved.

Kennett and Bailey stayed at the cabin with Gard Sawyers, the trapper. Lt. Clark headed back to Smoyer with Elvin Hess. The two of them carried the sleeping bag, a high altitude flying suit, an ax, a bottle of wine, and one half of the K-rations that the PBY had dropped. Lt. Clark, you will remember, had only one shoe. He wrapped his shoeless foot in a towel from the cabin, and went ahead anyhow. When I think of Lt. Clark, smoking his pipe in the top of the tree, climbing down on the pin knots, finding the can of carbide and knowing what to do with it - and now, heading six miles down Lake Creek with a load, in the rain, almost dark, and with only one shoe, I have a lot of respect for those Tennessee guys!

By now it was well after dark and Hess was exhausted. Fortunately, they met one of the Army searchers, Lt. Cummings, who was able to take part of Hess's load. When they finally got back to Smoyer, Harold Noah and Lt. DeMers along with two other soldiers were already there. One of the soldiers was a medic. According to the newspaper account later, he was identified as "a Negro named Stokes." He stayed with Smoyer for the next thirty-six hours, straight through, without sleep. It was probably through his efforts that Smoyer was able to pull through as well as he did.

When they got back to Smoyer they put a piece of malted chocolate from the K-rations into his mouth, put the high altitude flying suit on him, and placed him in the sleeping bag. None of the others had a sleeping bag. For them it was another miserable night.

Saturday dawned with a little less rain. It was a frustrating day. Elvin Hess got up early and went back to the cabin and about 10 a. m. the PBY returned. Elvin asked by walkie-talkie for blood plasma, sleeping bags, food and supplies. He also asked for loggers to cut a trail so that Smoyer could be carried out by stretcher. The material was dropped, but could not be retrieved.

Frustrated, Elvin, along with Flight Officer Kennett and Corporal Bailey, headed back to Smoyer. Kennett limped along with his bad ankle, and they arrived about 2:30. At 8 p. m. an Army doctor and about twenty soldiers arrived. They cut wood and made camp, but for the second night there was no food.

Elvin Hess went hunting but had no luck. All of the activity must have scared the elk out of the country. While he was out it got dark and he heard a big racket which he took to be a herd of elk. It turned out to be twenty-five loggers crashing through the brush in the dark, with two flashlights among them.

Hess led them off toward Smoyer and the camp, but two hours later they arrived back where they had started - they had gone in a circle. Hess then tried navigating with a compass, and finally got them through. There were a total of 10 K-rations to be divided among fifty men. Still, there was no bedding. Another horribly miserable night. Willis Peterson told me that he stood up all night on the lee side of a tree and was soaked.

Sunday morning at 6:30 they started back with Smoyer. This was the seventh day since he had pre-flighted the C-46D at Sedalia, Missouri. Afterward, he didn't remember much of the trip by stretcher.

The loggers broke trail and the soldiers carried the stretcher. Kennett's ankle was giving him a lot of trouble, and it was slow going. The soldiers managed the stretcher for about four and a half miles and gave up. The loggers took it the last mile and a half. Willis Peterson was one of the loggers. He told me that he wore all the hide off his hands on that stretcher. "That guy was in bad shape," he said. About a half mile from the advance camp they had to cross Lake Creek, with the stretcher, on a fallen log. This was the most dangerous maneuver of all, but they made it across.

At 12:45 Sunday afternoon the fifty of them staggered with Smoyer into the advance camp. There was coffee and hot food there. Earlier in the morning the aircraft had been found. It was on Pheasant Creek, only a couple miles away. The pilot, Captain Hugh B. McMillan of Kansas City and the co-pilot, Harlow A. Marsh of Syracuse, New York, were dead. Jerry Baughman, who was there when they found it, told me, "They were just starting to get ripe." Jerry rode out with them in the ambulance.

The crash site looked as though the pilot had tried to crash-land on the top of the trees. He didn't have much choice, of course, and might have made it but two of the trees uprooted, causing the plane to nose straight down the last 200 feet between some larger trees. The tail stood straight up, and the plane had to be pulled over with a cat to get at the bodies of the pilot and co-pilot. Jerry picked up a souvenir - a bent propellor blade. He gave it to me and I kept it in my bedroom for several years. Teen-age boys like things of that kind!

C-46D crash, nose down in the woods. Pilot and Copilot were dead, in their seats.

The rescuers and Smoyer were loaded onto some jeeps and weapons carriers, which were pushed with three bulldozers through the mud for most of the fourteen miles back to Cle Wilkinson's place. They arrived there at about 3:30 p. m. An ambulance took Smoyer to town and he arrived at the McAuley Hospital in Coos Bay at 6:45 p. m., 170 hours after he bailed out. It had been one hell of an experience. "I was very sick for about a week and I was starved," Smoyer told me in a letter.

"They took very good care of me at McAuley Hospital and then I was transferred to Portland Army Hospital." He had gangrene in his feet, the final stage of immersion foot syndrome, which we see in too many of the rescue stories. He remained in Army hospitals for about nine months before being discharged from the Air Force in August 1946. His feet were saved, but it was several years before he fully recovered He returned to his pre-war job at Bethlehem Steel where he remained until retirement.

The radio operator, Sergeant Robert T. Neal, was never found. I suspect that his chute may not have opened, and he landed in the raging waters of Lake Creek. Otherwise, one would think some trace would have been found. During the search, his fiancee's mother called the *Coos Bay Times* to say that she had no word from him since he called the night before they left Oakland, California.

Harold Noah, Sam Crawford, and Alfred Leaton, had remained in the woods, searching for Sgt Neal's remains, for a week after the search was called off by the Army. On December 13 Neal's parents arrived in Coos Bay from their home in Los Angeles but Harold Noah told them that in his opinion further search is futile "under the present conditions." He intended to return to the area after the water subsided, and search along the stream.[2]

Advanced Command Post near Lake Creek

Neal's parents returned each summer for several years and hired guides to search the area for clues. His mother, who didn't have hiking boots, would borrow boots from Pat Wilkinson for those trips. Several years after the crash Neal's mother hired Elvin Hess to do a final search. Elvin didn't want to do it, but he was afraid that if someone else did it they wouldn't do it right. He spent about two weeks thoroughly searching the area. He didn't find anything and that was the last attempt to solve the mystery of what had happened to Neal.

I don't know what happened afterward to the other survivors. Each of their stories would be a chapter in the social history of post-war America, I'm sure. If the resourcefulness they showed in the woods above the falls is any indication of what they were able to do later, we can assume that they did all right. I sure hope they did.

Of the rescuers, most remained in the woods in Coos County for the rest of their working lives. This was certainly true for Archie Clawson, Harold Noah, Elvin Hess, Roy Spires, and many others. Harold Noah was involved in another spectacular rescue

---

[2] See *Coos Bay Times* Dec 10 and 13, 1945. "Parents of Missing Flyer Plan Visit," and "Searcher Says Effort Futile."

on the upper West Fork in 1949. A hunter had injured himself and was lost and given up for dead. Harold Noah didn't give up, and found him alive after a week in the woods (see chapter 5).

Roy Spires, the high climber who rescued David Reid from his perch 150 feet in the old-growth tree, continued in the woods. I was told that when he was eighty years old he gave a high-climbing demonstration at the Coos County Fair. He was from an old pioneer family of Myrtle Point, and had started in the woods back in the steam donkey days before World War I. He was the stuff of legend.[3]

During the 1960's many of the loggers from Coos County moved to Alaska. Elvin Hess was one of them. While there, he was cruising timber and ran across a Kodiak bear who grabbed him, shook him unconscious, and walked away. Elvin woke up just as the bear ambled out of sight. With his scalp ripped open and one eye hanging alongside his head, Elvin crawled out of the woods to a road where he was eventually picked up by a passing motorist. He was delivered to an airplane and flown to Juneau, where he remained in critical condition at the hospital for several weeks. Within a year Elvin was back working in the woods, where he remained the rest of this life.

As for the land above the falls, it would never be the same again. The world knew about it now. There had been more men involved in the search and rescue of the flyers than had ever been in those woods before, from the beginning. Newspaper and radio coverage emphasized that this was "one of the wildest, most thickly forested sections of the state." It caught the consciousness of a public only beginning to be aware that wilderness was becoming a thing of the past.

In 1948 Weyerhaeuser contracted with the Morrison-Knudson Corporation of Boise, Idaho, to build the major logging road from Allegany up the East Fork and over the Matson Creek Falls. This road ultimately connected into the upper Glenn Creek and upper Lake Creek areas.

In July 1950, Weyerhaeuser hauled the first load of logs out of this forest. The logging started on Tyberg (Conklin) Creek just above the Matson Creek Falls. Fifteen years later they were into Lake Creek, hauling logs out of the area where the C-46 had crashed. Remains of parachutes were found, but no sign of any remains of Sergeant Neal.

As far as I know, none of the rescuers received any compensation for their efforts. They hadn't expected any. No awards or medals were given for what, by any standards, were heroic actions. None were expected. The stories about the rescue were told and

---

[3] Roy Spires' daughter Kathy Larsen told me that a month later her father returned to the tree from which he had rescued the flier David Reid. With his belt and spurs Roy climbed to the crown of the tree, retrieved the parachute, and took it home. Kathy told me that as a child she had clothing made from the parachute material and to this day she has the parachute straps as a souvenir. Interview Feb. 27, 2011.

re-told in taverns and bunkhouses for sixty years but at this late date I can paraphrase Longfellow by saying, "Twas the end of November in forty-five but hardly a man is still alive, who remembers that famous day and year." Perhaps by having set the story down, in writing, it will live on for another generation.

High Climber Roy Spires, starting up a tree with his belt and spurs, 1945.

Roy Spires, right, in a spar tree, about 1945. He is standing about 175 feet in the air. There would have originally been a crown in the tree which went up another 40 feet or so. That is where the flyer David Reid was hung for 36 hours before being rescued by Roy Spires.

Appendix to Chapter 4

A letter from Walter A. Smoyer

Hellertown, Pa

December, 1995
Dear Lionel,

We took off from Sedalia, Mo. on Nov. 24, 1945. One of our planes was grounded in Oakland, California and we were going to pick up the four pilots and take them to Portland, Oregon. We stopped in Denver, Colorado for fuel and picked up four more passengers going west. From Denver we went to Ogden, Utah. We flew over seven hours and I was plenty tired. The next day (25th) we had a rough trip. Then we flew from Ogden to Fresno, California over the Rocky Mts. We flew at 16,000 ft. Since we had no oxygen I felt pretty sick after 2 hrs. of flying. We flew up to Oakland, California and stayed for the night. Sgt. Neal and I went to San Francisco across the Bay from Oakland and since Neal had been there before he showed me around. We had a great time riding the Cable Cars and then took a Ferry ride across the Bay. The whole evening cost me $1.05 and we had a great time.

The next morning (26th) we picked the other four passengers and left for Portland, Oregon. We had 12 men on board – four of the crew and 8 officers. We had flown about 4 hrs. when things began to happen. The weather grew bad and the plane began to collect ice. The storm grew worse and our radio went out. We were by this time near Portland but because of the storm we could not contact the ground. The Pilot decided that we had better turn back and try to make California. We started back but ended up lost and almost out of gas. We had no radio and we could not see the ground, so we did not know if we were over land or water. We had flown about 6 hours when the Pilot gave orders to put our chutes on and get ready to bail out as soon as we sighted land and by this time one of our engines cut out because we used up all the gas.

The cargo door was opened and we started to jump. The officers jumped first and after that it was Sgt. Neal's turn. He was very scared. He kept asking me if he should count to four before pulling the rip chord. I told him to pull it right away as we were very low. I went out last and my chute opened and I started to sail to the ground. My chute caught on the top of a Giant fir tree that must have been 250 ft. high. I looked at my watch and I landed in the tree at 4 p. m. I heard the plane crash and then everything was silent. I do not know why the Pilot and co-pilot did not bail out as they would have had enough time to get out.

There I was all alone just hanging in my chute and scared to death that my chute would blow off the tree. The wind blew me into the trunk of the tree. I tied myself to the trunk of the tree that night. I don't think I slept but I did doze off a little. The storm lasted all night and by Tuesday morning I was soaked thru and thru from the rain.

Tuesday it took me about 4 hours to get out of the tree. I climbed down about 100 ft. and then I had to jump the rest which must have been about 50 ft. The jump knocked me out for about an hour. When I came to I was very lucky as I did not break any bones and I could walk. I walked for 6 days in the wildest country I have ever seen or ever hope to see again. I had nothing to eat but I had some chewing gum and there was plenty of water to drink. I don't ever remember being hungry.

The search party found me on Saturday and I was just about gone by that time from exposure and hunger. They gave me 4 pts. of blood plasma and shots of morphine. I was kept in the woods until Sunday until they could get a stretcher in. The place was so wild that they had to cut a path to get me out. They took me to Coos Bay Hospital which was about 38 miles from where they found me and the nearest town. I don't remember too much after I was found.

I was very sick for about a week and I was starved. They took very good care of me at McAuley Hospital and then I was transferred to Portland Army Hospital. I was told I would be there until I was well enough to travel. I could not walk for about 5 months. I had no feeling in my feet and my legs were beat up from the knees down. I had no pain but my feet felt as though they were asleep. I just walked on them until they were numb. They were all black and blue and swollen and later I found out I had gangrene.

Sometime later I was transferred to an Army Hospital in Spokane, Washington for rehabilitation. My wife contacted her Congressman and asked to have me sent east and closer to home. So in April I was sent cross-country by train in a private car and a male nurse to attend to me. I was sent to Camp Upton, Long Island, N. Y. I was only there a short time and then transferred to Walter Reed Hospital in Washington D. C. I stayed there until I was discharged from the Air Force in Aug. 1946. I then went back to Bethlehem, Pa, where my wife and 3 yr. old son were living with her parents.

For many years after the crash I had psychological effects in the month of November. I would just shiver and actually feel sick.

I took this information from a letter I wrote to my wife on Dec. 8, 1945 as the crash was still very clear in my mind.

Thank you for taking such an interest in my story and I sure was a lucky fellow.

Sincerely,

Walter A. Smoyer

Extracts of letters from Mildred Smoyer

Nov. 16, 1995
Dear Mr. Youst,

After all these years a man by the name of John Anderson contacted my husband. Mr. Anderson was in the search party & helped to rescue my husband. He told us about a book which was written by you and that there is an Epilogue about the crash in the book.

The reason I am writing to you is to find out if this book is still available. We would love to purchase this book to pass down to our grand-children.

I also thought you might like a follow-up as to what happened to my husband after he was found. He spent nearly a year in Veteran's Hospitals after which he was discharged in 1946. He developed gangrene on both feet and for awhile the doctors thought he might lose them. God was good to us and his feet were saved. Walter went back to his job at the Steel Mills and spent 30 years there. He retired in 1982 and we have been married 52 yrs. We also have 3 children and 3 Grand-children.

We would appreciate hearing from you.

Sincerely,

Mildred Smoyer
Hellertown, Pa

Nov. 29, 1995
Dear Mr. Youst,

We received your most welcome letter and book. We read with great interest the account of the plane crash and all the men who survived the crash. Also all the brave and courageous people who were involved in the search. If it weren't for them my life would have been a different story. We shall always be indebted to all those people whom we never met and will never be able to thank.

Again may I thank you for writing and only wish I could convey my thanks to every one who was involved.

Sincerely,
Mildred Smoyer

Extracts of a letter from John R. Anderson

October 31, 1994

Dear Mr. Youst:

At the time I was in a Search and Rescue Unit as master of a '42' crash boat. We were stationed on the Columbia River at the end of the most used runway at Portland A F Base (between Hayden Island and Government Island). When the plane crash at Coos Bay occurred we were called there for the search. We came in on Wednesday.

I was with the group that was with Pfc Smoyer and helped on the long carry down. Only since I got your book did I learn that he made a fast recovery. Most of us, of course, were completely out of our usual environment. None of us had been in a wilderness mountain forest before, but we were young and tried to adapt the best we could and give a good account of ourselves. The "spiked" boots that we got were a great help. Without them I fear we would have had more injuries.

I was a bit disappointed in that no mention was made, in your book, of the Portland AF Base Unit at all. In fact, I am 99% certain that I see myself and my Lt. in the photo captioned "Soldiers from Camp Adair etc" on page 297. We all looked alike, however, wet and cold. No big deal.

Sincerely,
John R. Anderson

A portfolio of pictures of the search and rescue from the album of Alice Wilkinson Allen

Bodies of the pilot and copilot had not yet been removed.

Note open hatch through wich the men had bailed out.

Makeshift bridge

Jeeps heading for Lake Creek

The forward command post, Lake Creek

Lake Creek Falls, Ash, Oregon

Below is the old Joe Schapers homestead above the Golden Falls of Glenn Creek, November 1945. Joe Schapers' step-son-in-law Cle Wilkinson and Cle's brother-in-law Alfred Leaton were living there at the time of the search and rescue of the Army C-46 aircraft. The ranch became the principal headquarters for the search because it was the last place that cars could drive to, and radio communication with town was possible from there.

Cle Wilkinson's house, left. Barn, right. Note ambulance in front of the barn. Photo by Gerald Baughman, who directed the ambulance to the site.

Note on pictures and sources: Photos of the rescued flyers were taken at the McAuley Hospital in Coos Bay by the *Coos Bay Harbor* newspaper, of North Bend. The photos are now in the collection of the Coos County Historical Museum and are reproduced here with grateful appreciation. The other photos are from the albums of Alice Allen, Audrey Milton, and Jerry Baughman. I have talked casually and otherwise with the following persons, and have incorporated information from them in various parts of the story: Hank Gosney, Alice Allen, Gary Hess, Willis Peterson, Dow Beckham, Jerry Baughman, and Kathy Spires Larsen. The contemporary newspaper accounts in the *Coos Bay Times* by reporter Leroy Inman and Editor Walter J. Coover were excellent and form the basis of what I have assembled here. I am also grateful to Dow Beckham for allowing me to use the unpublished manuscript of an article he once wrote about the rescue. Following the original publication of *Above the Falls* in 1992, some new information has come to my attention. Coos Bay *World* reporter Elise Hammer did interviews with Walter Smoyer and with Archie Clawson in 1995, and her article appeared in the November 27, 1995 edition of the paper. I was able to exchange letters with rescued flyer Walter Smoyer of Bethlehem, Pennsylvania, and his wife Mildred, and with one of the soldiers on the rescue, John P. Anderson of Milford, Kansas. From their information I was able to correct certain errors in the original, and add a few more details.

Map VI

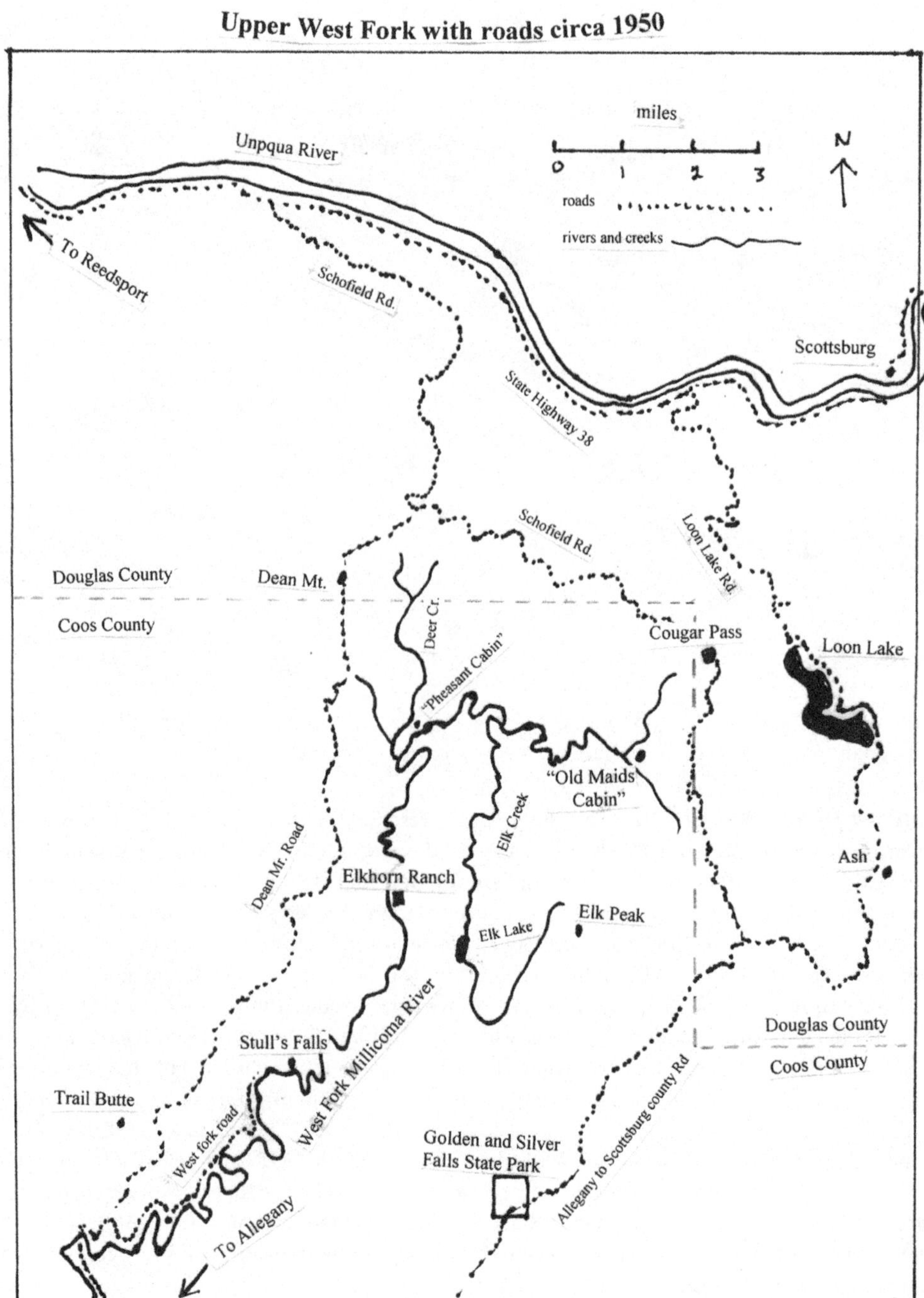
Upper West Fork with roads circa 1950
miles
0 1 2 3
N
roads
rivers and creeks
Unpqua River
To Reedsport
Schofield Rd.
State Highway 38
Scottsburg
Schofield Rd.
Loon Lake Rd.
Douglas County
Coos County
Dean Mt.
Deer Cr.
Cougar Pass
Loon Lake
"Pheasant Cabin"
"Old Maids Cabin"
Elk Creek
Ash
Dean Mt. Road
Elkhorn Ranch
Elk Lake
Elk Peak
West Fork Millicoma River
Stull's Falls
Douglas County
Coos County
Trail Butte
West fork road
Golden and Silver Falls State Park
Allegany to Scottsburg county Rd
To Allegany

Chapter 5

An Elk Hunt Gone Wrong

(The search and rescue of a story)

Getting the facts to write this account was a long time coming. It is the case of a memorable event becoming lost to the collective memory, and of its recovery. To begin at the beginning, I had been in the Air Force overseas for several years and completely out of touch with what was going on here at home. I didn't know about the Columbus Day Storm of 1962 (I was in Libya at that time), nor had I heard of the system of logging roads that had been built in the aftermath of that storm.

In 1967 I took a 10-day leave to visit my dad in Coos Bay, and one of the things he wanted to show me was the new logging roads in the Elliott State Forest, which opened a hundred square miles of territory I had never seen. We drove from Lakeside up Benson Creek Road to its junction with the Dean Mountain Road and south on it a few miles where we turned east, down Trout Creek to the West Fork, across it on the new concrete bridge and up Beaver Creek to the 1000 Line (the Elk Ridge Road) and down into Marlow Creek, finally coming out at Allegany. I'm looking at an Elliott State Forest map as I write this, and it was about 30 miles, the way we went. But we had no map that day and we had no clue where we would be coming out. My "mental map" as we drove would have brought us out on Schofield Creek near Reedsport, if it had been right. But I was completely turned around and could not have been more surprised than I was when we came out on Marlow Creek at Allegany! It was a great adventure for me and my dad and the first time I had seen so much of the back country so near to where I grew up – and I learned how easy it is to get turned around up there.

During the drive, we talked. I hadn't been home in years and there was a lot to catch up on. Dad was a fine story teller and I listened while he brought me up to date. As we descended Trout Creek into the West Fork he was reminded of the story of a lost hunter. There had been quite a search for him, but after a few days it was called off and declared hopeless. Dad said there was a logger who lived at Eastside – when the carnival was held at the old Eastside airport would wrestle for purse money and did pretty well at it. His name, I later found out, was Harold Noah.[1] After the search was called off, Harold decided that he could find that guy. According to the story as my dad told it, Harold and a friend took a small rubber raft, going down from the headwaters of the West Fork, sometimes floating, sometimes towing that rubber raft through the boulders and windfalls in the river, which was up dangerously from the recent rains. They stopped and called out every few minutes and, according to the story as Dad told it, at one point they heard a tapping on the bank above them. They climbed up and found the injured man with a broken leg lying under the dry side of a log, too weak to talk but tapping the log with a rock he held in one hand. That is the story as I heard it from my dad during that drive through the Elliott in 1967 and it made a profound impression on me. Except for the part about the rock and the log and the broken leg, it all checked out when I finally got the documented facts.

After that, I heard mention of the story from time to time, but never bothered to get any further details. Benson Judy, who was involved with a part of the search, made reference to it when I interviewed him about something else in 1993. So did Hershel Minkler – whom I interviewed at great length in 2003 for the Coos County Oral History Project. Wilma Lund, granddaughter of Fred Noah, mentioned it, as did Alice Allen, daughter of Cle Wilkinson, and others. Nobody ever said when it happened. It became something like a myth that was outside of time – or "once upon a time." It was an event in the local past that had no need of being anchored to the calendar. Usually when the subject came up in conversation, it was with people who were quite elderly and are now deceased. I was completely ignorant as to when this event might have taken place, but I did have the name of Harold Noah as the rescuer, and I had an outline of the story.

Fast forward to the writing of this book. As I collected search-and-rescue stories of the Elliott and the Weyerhaeuser timberlands it struck me that to make the cycle complete, I needed the story of Harold Noah and his rescue of the lost hunter on the West Fork. But by now everybody that I knew who remembered anything about it was dead, and so I began to search for the story of the search, and that is a story in itself.

The Noah's are all descended from the pioneer John Wesley Noah, who moved his family into the East Fork of the Millicoma back in 1884. He and his wife Mary had 15 children, 14 of whom lived long enough to have children of their own, and so the Noah family has

[1] I remember in probably 1947 I was at the carnival and witnessed the professional wrestler drumming up challengers. For a price, anyone could enter the ring and win the prize money by pinning the professional wrestler to the count of one. For the professional to win he had to pin the challenger to the count of three, which seldom took him more than a minute or two. I am sure it was Harold Noah whom I saw enter the ring and pin the professional, the only challanger to do so during that afternoon. Harold Noah was a big man, and he was stout as a bull.

many lines and branches all through the area. The parents of Harold and his brother Leonard were John and Emma Noah, who once lived on the Charles Crane place on Glenn Creek, and it is presumed that Harold lived there when he was young.[2] The census and other records show that Harold was born in Coos County on February 10, 1912 and died at Grande Ronde, Polk County, in March 1981. He married Nora Crawford of Broadbent (near Myrtle Point) in 1929 when he was 17 and she was 18 years old. They had five daughters over their many years together. Harold's brother Leonard married Nora's sister Betty Crawford and they had five sons. This is information I received in a conversation with Wally Noah, who has a pretty good grasp on the family history. He couldn't remember any details about Harold Noah's search and rescue of the lost hunter, nor when it happened, but he suggested that I talk to Jimmy Leeth, whose mother was Amanda Noah. He is about 86, Wally told me, and has a perfectly good memory.

I did talk to Jimmy Leeth, and he was a wealth of information but not even he could remember when the search and rescue occurred. He said that the rescued man had a "frozen leg" and that he "gave Harold his house for saving his life," and that it was "Babe" [Leland] Page who was with Harold on the rescue. Babe was Harold's hunting partner and they hunted together for years, according to Jimmy. Babe worked for C. J. O'Neil's Radiators most of his life and Jim O'Neil should know something about it and, Jimmy added, Harold Noah was a Mason. He gave me the name of an elderly Mason who might know something about it, but when I talked to the old guy, he didn't know who I was talking about. Jim O'Neil didn't remember the event either, but he said that he got out of the Army in 1947, and so it had to be before that or he would have known about it.

The year 1947 at least gave me something to start on. I went to the microfilm files of the daily *Coos Bay Times*, and I searched the hunting seasons (October and November) for the years from 1947 all the way back to 1940. Harold Noah's name came up one time, but that was in regard to the Army C-46 aircraft that crashed on the Weyerhaeuser lands in November 1945. Harold Noah, his brother-in-law Sam Crawford, and Alfred Leaton stayed out in the woods for three weeks after the search had been called off, looking for traces of radio operator Sgt. Robert Neal, whose body had not been found. Harold Noah was quoted as saying that the body would probably never be found, that his chute had probably not opened and he probably went into Lake Creek, which was at a "raging torrent".[3] That is the consensus of opinion to this day.

---

[2] Alice Allen remembered that "they lived on the property, not in the Crane house." Interview Sept. 25, 2010.

[3] *Coos Bay Times*, Dec. 13, 1945, p.1, "Searcher Says Effort Futile."

Going through all those years of hunting seasons was quite informative, however, and not a waste of time. The year 1940 was the beginning of the modern period of elk hunting in Coos County. The State lifted its moratorium on elk hunting that year, the first open season for elk on the Coast Range since 1911. During the 1940 season the rules required "parties of at least two hunters, provided with blocks and tackles, or sufficient rope to hang the beasts for butchering . . .". This was because when the season had opened in 1933 in Northeastern Oregon, there were thousands of hunters but "hundreds of elk were slain and left to rot when their hunters found themselves unable to handle the huge animals."[4] Most of the hunters had experience with deer that dressed out at well under 100 pounds, but had no experience in dressing out elk, which often went to over 700 pounds. One man alone could not handle such a large animal. Warren Browning remembered the 1940 season, saying, "God, everybody went mad, hunting. The state must have made a billion dollars sellin' elk tags! It hadn't been open for years, you know. Lots of elk."[5] That was the year that George Rice, who had hunted every year since he was old enough to hold a gun, got his first legal elk. Then came WW II and most men of hunting age were either in the service or working in war industries some place. Gasoline was severely rationed, and ammunition was unavailable. There was virtually no sports hunting done during those war years, but as soon as the war was over, everybody again went mad, hunting!

Elk killed by William C. Morgan, 2 September 1941. 6-point, Weight, 890 lbs. Pictured, l to r: George Rice, Kirby Brunfield, and William C. Morgan. Others in the hunting party were Kenneth Smith and Philip R. Morgan. Killed with 300 Savage at approximately 150 yards with one shot in the right shoulder. Photo was taken at Charlie Feller's fish plant on Front Street in Marshfield. Photo courtesy Lynne and Dennis Rice. © Lynne Rice, 2010.

---

4 *Coos Bay Times*, Nov. 7, 1940, p. 13.

5 From my interview with Warren Browning, February 11, 1992. The entire interview is in *Above The Falls, 2nd Edition*, pp. 217-227.

The war ended in August 1945, and that hunting season the newspaper was full of news about lost hunters, injured hunters, and hunters killed from stray bullets. That trend continued through the rest of the '40's. Typical items: "Reedsport Man Shoots Hunter," "Hunter Drowned in Coos River," "Portland Man is Hunting Victim," "Two Women Hunters Reported Missing," "Hunter's Body Found Sunday," "Two hunters, both about 60 years old, are missing in the hills near Golden and Silver falls."[6] The first weeekend of the 1948 season, "Deer Hunting Takes 6 Lives." The following week the headline stated, "Hunter Deaths Climb to 17."[7] The casualty rate was so high it was as though the war had never ended!

The newspaper seems to have noted every person reported as lost or injured, and so it appeared certain that the Harold Noah rescue of the lost hunter would be in the paper, if I could find it. It did not appear during any of the hunting seasons of the 1940's. I asked Alice Allen, daughter of Cle Wilkinson, if she had any idea when it might have happened. She thought a minute, then said, "It was after 1947. We were married in 1947 and it was after we were married. It must have been in the late 1950's or early 1960's." The man's last name was Carroll, she said, but she didn't remember his first name. OK – that was a real help. I went to the newspaper microfilm files, and covered all the hunting seasons between 1954 and 1961. It took a little more than an hour to do October and November of each year, and so by then I had quite a bit invested in time. There was nothing about the Harold Noah rescue of the lost hunter, but plenty more about the hunting seasons, rescues, injuries, and deaths. Someone suggested that I ask Loon Lake Lloyd (Lloyd Keeland), because he knew everything that happened around Loon Lake. I called him – it was the day after his 90th birthday. He did not hesitate. "It was the year I came to Loon Lake," he said, "1953." I looked at the newspaper microfilm for 1953, and it wasn't there either.

Needless to say, I was getting quite discouraged. There were only four years remaining for me to look at. I planned to look at 1962 and 3, and then go back to 1950–51, which I had skipped. It was not in 1962. I went to 1950 and the front page of the November 2 issue had the first account: "Coos Bay Elk Hunter Sought. A 58-year-old Coos Bay man, Clayton Carroll of 719 south Second Street, is the object of an intensive search in the lonely Cougar Pass area on the Coos-Douglas county line east and south of Reedsport." He had gone alone into the area with his jeep and camping gear on Tuesday, October 24, in preparation for the

---

[6] See *Coos Bay Times*, p. 1, Oct 17 and 22, 1945; Oct 1, 5, and Nov 18, 1946; Oct 28, 1947 for these examples.

[7] See *Coos Bay Times*, p. 1, Oct 4, and 12, 1948.

opening of elk season the following day.[8] Eureka!! My search was successful. Now I could rescue the story!

Two years earlier Clayton Carroll had moved his family to Coos Bay from Colorado. During that time (1948–1950) work was plentiful around Coos Bay, and large numbers of men from the Midwest and the South had arrived to work in the scores of small sawmills that were operating in the timberlands throughout the county. Carroll was working for Coast Pacific Lumber Company, which operated a planer mill at the Portland Dock in Bunker Hill, where many of the small sawmills delivered their rough lumber for finishing. Hunting season was such an important event in the life of many loggers and sawmill workers that it was not unusual for a mill to close down so that the entire crew could go hunting if it wanted to. That may explain how Carroll happened to be off in the middle of the week.

Information in the article gave names and specifics that allowed me to find obituaries that gave more names and one of those names was that of his daughter Patricia Wollin.[9] I looked in the telephone book and she was there. I was on a roll, as far as getting the story together and when I called her, she gave me several new details that helped to fill it in. She said that her father was an avid hunter all his life, and very much an outdoorsman. She said that he hunted every October in the Rockies when they lived in Colorado. She remembered that it always snowed, and he always missed her birthday. This was the first time he had ever gone out alone, she said. Comparatively new to the area, he may not have been able to find a hunting partner and that may explain why he was by himself.

According to the newspaper accounts, he had stopped at the State Police office in Reedsport to get information on the road into the Elliott, and to let them know he would be up there. At that time there were only about 27 miles of road in the Elliott State Forest, all of it built between 1933 and 1942 by the Civilian Conservation Corp (CCC). The Schofield Ridge Road ran about 15 miles from Reedsport to the Cougar Pass Lookout, following the old Forest Service trail. The CCC had blasted and bulldozed it into a road for motorized vehicles and extended it on to Loon Lake. Carroll set up his camp at the Cougar Pass Lookout on Tuesday, October 24. The next morning elk hunting season opened and he drove his jeep about two miles to where Gould's old hunting trail crossed the road. He parked the jeep there and went down Gould's trail to the headwaters of the West Fork of the Millicoma River. The

[8] The event occurred in October and November, 1950. During that time I was working at the G. L. Spiers logging camp on the Bald Hills east of Orick, California. That explains why I was so unaware of when it might have happened.

[9] I am especially grateful to Lynne Rice, who provided me the burial information on Clayton Carroll, which in turn led me to his obituary and to more family information.

Cougar Pass Lookout, 1941. Photo courtesy Alice Wilkinson Allen.

next day, Thursday, October 26, other elk hunters saw him near the "Old Maid's Cabin," which was on the West Fork only about a mile from where he had parked his jeep.[10] "This is the last time that anyone has reported seeing the missing man" according the the state police officer James Joy, who headed up the search activity once it got started a week later.[11]

The weather was the problem. There was occasional rain on the 24th and 25th, which was good for the opening of elk season that year. Temperatures held in the 50's, day and night. But on October 26, gale warnings were given for winds up to 75 mph that afternoon. Barometric pressure had dropped to 29.7, portending a real storm early in the season. The storm, as it progressed into the next day, was reported to be 1000 miles long and 500 miles deep, with gale winds barely below hurricane force. On the 28th the *Coos Bay Times* reported that statewide "at least 11 lives reported lost since Thursday." The roof was blown off the Cape Arago sawmill at Empire. At Allegany a 2400-volt power line was blown onto one of Jess Ott's cows and killed her instantly. On Monday, Oct 30, the *Times* reported, "Coos County has one person known dead and one other believed dead as a result of a severe storm that battered southwestern Oregon over the weekend." That same day, state game officer Richard Miles found Carroll's camp at Cougar Pass, "complete with sleeping bag, stove and groceries."

According to Carroll's daughter Patricia, he had shot at an elk and wounded it. While following the elk to finish it off, the fog and mist which came

---

[10] George Gould and his family had several camps in the old Coos Bay burn, which were used as headquarters when hunting elk and deer for market. From *The Goulds of Elkhorn*, pp. 40-1, "An idea of the amount of meat can be formed from the fact that at one time George and I killed five elk weighing from 400 to 700 pounds each." They could smoke the meat from 25 to 30 deer at one time, before packing it out to Allegany for shipment to San Francisco. The "Old Maid's Cabin" was one of the hunting camps, at the mouth of Cougar Creek west of Cougar Pass. The "Old Maid" was Olea McClay, an unmarried sister-in-law of George Gould and teacher of his children. see also, Jerry Phillips, pp. 375-6.

[11] *Coos Bay Times*, November 3, 1950.

ahead of the storm rolled in around him and he lost his bearings. Then came the storm itself, which lasted almost a week. Carroll was wearing rubber boots and before long he was suffering from immersion foot (trench foot). He used his gun as a crutch for as long as he could, but he was soon unable to stand or walk. It was not a "broken leg," as my dad thought it was, nor was it a "frozen leg," as Jimmy Leeth stated. It was immersion foot, which can be every bit as debilitating as either a broken or a frozen leg. He had evidently found the trace of Gould's hunting trail along the West Fork and was heading downriver. Few hunters would ever admit to being "lost," and I have been corrected when using the term in connection with this story.[12] Carroll was an "injured" hunter, lost only to the men who were searching for him.

There were many large snags in the Elliott, silent reminders of the old-growth forest that stood there prior to the great Coos Bay fire of 1868. Carroll was able to burrow into the rotten side of one of those snags, making a kind of cave that kept him out of the weather a little bit. It was about 300 feet above the river, not far from the George Gould hunting trail. He could not stand and walk, but he could crawl down to the river for water, where he left a visible trail. His daughter said that he ate a few berries (there would have been black huckleberries ripe at that time), and he ate some leaves, and that he kept a rock in his mouth to help assuage his hunger. He was probably there from October 26 or 27 until he was found on November 4. Later, at the hospital, he said, "At night when I would lie under a snag a little furry animal would come and cuddle up close to my collar and then burrow under my chin. I never did know what it was – I guess I was too tired to even try to guess."[13]

On Tuesday, October 31, the newspaper reported, "Two more drowned in Coos Flood Waters as Rescue Boat Swamps Nearing Safety." That day Carroll's son-in-law James Wollin notified the police in Reedsport that Carroll had not come out. By Wednesday the weather was getting back to normal and Wollin organized a search party from men in Ash Valley and Loon Lake area, who began searching the following morning, Thursday, November 2.[14] On

[12] See Robert J. Koester, *Lost Person Behavior*, page 193: "Hunters' focus/pursuit of game is the primary cause of becoming lost. They go to great lengths to self-rescue. Many hunters will avoid using the word "lost" when confronted with searchers."

[13] *Coos Bay Times*, Nov 7, 1950.

[14] Pierce Duckett of Loon Lake compiled a list of men who aided in the search. He listed C. L. Leggett, Benson Judy, Marvin Judy, Herschel Minkler, Charley Thompkinson, Loren Thompkinson, Ross Johnson, Harold Parker, Joe Muskrat, Marvin Haskel, Leonard Gustafson, and Elmer Nelson. State trooper James Joy added Carl Collins, George Clemmons, Pierce Duckett, Harold Dick, Ed Tolman, Everett Smith, E. O. Williams, E. I. Williams, Charley Thornton, Loren Thornton, Bud Blondell, Raz Johnson, Dave Waler, Buster McCrary, and Harold Noah. *Coos Bay Times*, Nov. 6, 1950, p. 1.

that day for the first time, we read in the paper about the "Coos Bay Elk Hunter Sought." Fifty-eight-year-old Clayton Carroll had been out in that storm for a week, from its beginning to its end. On Friday, November 3, the paper reported that Carroll had last been seen a week earlier, on Thursday, October 26 – a day after hunting season opened. On Saturday, November 4, the State Police called off the search, after only two days. Officer James Joy announced, "The area has been covered as thoroughly as possible."[15] Carroll's wife Hester evidently took issue with that assertion. She offered a $500 reward for anyone finding her husband because, as their daughter Patricia told me, "We could accept that he was dead, but we couldn't accept that he couldn't be found."

Another person who thought that the area had not been covered as thoroughly as possible was the logger, wrestler and general woodsman Harold Noah. He had been on the search with the rest and had a very definite idea of where to look next. He knew the West Fork of the Millicoma like the back of his hand, having run traplines there during the 1930's. Now 38 years old, a timber faller since his teen years, and he had been on the 1945 search for the Army flyers who were lost on Lake Creek and stayed on the search for three weeks after it had been called off, looking for that last flyer. He had a good sense of where somebody could be found, and his idea was radical and bold. Instead of following the established trails, he would go into the headwaters of the West Fork from the road to Cougar Pass Lookout, taking with him a small inflated rubber boat. He would go down the river, stopping and yelling at intervals, coming out about 15 miles downstream at Stulls Falls, a little above where the road from Allegany terminated. Hopefully, he would have Carroll with him when he got there. He told the State Police in Reedsport of his plan, and asked that if he didn't come out at Stull's Falls within a reasonable time that help be sent into the woods "as we would have found our man."

Noah asked his long-time hunting partner Leland "Babe" Page if he would like to go along to help with the boat. Of course he would go. As Page told a reporter afterwards, "It was all Noah's idea. I went along because he needed someone to handle the boat. Noah knows that country very well and he figured Carroll was close to the water."[16]

---

[15] Usually it is the Sheriff's Department that is responsible for search and rescue. In this case the State Police took the lead probably because it was in the State forest. There was no office for the Elliott State Forest at that time. Also, Carroll's camp was barely on the Douglas County side of the line, but he was lost on the Coos County side. It may have been that the State Police helped to avoid a jurisdictional difficulty between the sheriff's departments of the two counties.

[16] *Coos Bay Times*, Nov 6, 1950. Quotations from the next three paragraphs are also from that article.

Early Friday morning, November 3rd (the day before Mrs. Carroll announced the reward), Harold Noah and Babe Page arrived at Benson Judy's house above Loon Lake, in Ash Valley. Benson drove them in his Jeep up the Douglas County road toward Glenn Creek in Coos County, then to the top of Dean's Mountain Road where he left them with their rubber boat. Harold's confidence in his ability to negotiate the 15 miles of river to the roadhead at Stulls Falls above Allegany was sublime. He brought no food or other supplies, nothing to encumber them in getting themselves and the boat to wherever Clayton Carroll was, and then to get Carroll on down the river to civilization. From the road, they carried the boat two or three miles, until about 10 a. m. when they were far enough down the river to where "there was just enough water to float the boat." About five miles from where they started on the road, they saw a trail leading from the river up the bank on the north side. They called out and heard a faint reply.

Cle Wilkinson and Baldy Crane's "Pheasant Cabin," at mouth of Deer Creek, West Fork, Millicoma River. Photo by Jerry Phillips.

Baldy and Prescilla Crane. After he was widowed, Baldy and his childhood friend Cle Wilkinson trapped and picked ferns for years. Courtesy Howard Lorman.

They had been on the search for less than six hours and had found their man a mile below the mouth of Elk Creek, about 300 feet above the north side of the river. He was huddled at the base of a large snag, "having scraped out part of the rotten core to get a little protection from the weather." He was stiff, wet, sore and terribly hungry. He asked, "Have you got anything to eat?" The answer was no. They hadn't brought a scrap of food with them! "You can't pack me out and I can't walk," Carroll said. There was a boat nearby, he was told. They loaded him into it – he couldn't walk and his hands were so swollen that he couldn't hang on, and so they had to tie him in. They later told a reporter, "We nearly lost him two or three times going downstream." With the West Fork at flood stage following a week of tropical storms, this was a precarious operation by any standard. It took them three hours of careful negotiation to get him the mile and a half to the mouth of Deer Creek, the site of Cle Wilkinson and Baldy Crane's "Pheasant Cabin." This was the headquarters they used for their commercial picking of sword ferns and was stocked with food and sleeping bags.[17] They spent the night there and Harold and Babe no doubt ate quite heartily while they spoon fed the starving man a little gruel and light nourishment.

The next morning, Saturday, November 4th, they started out again, down a raging river in their little rubber boat. The swelling in Carroll's hands had gone down, so he could help hold onto the rope. The bottom was soon ripped out of the boat and they had left only the inflated rim, like a large inner tube. From Cle Wilkinson's Pheasant Cabin to Stulls Falls is about ten miles by the river. The three men, one of them too weak to take care of himself, floating in the cold water most of the day, makes one wonder: what gave these men such strength of will? About four miles down, they passed the deserted buildings of the Elkhorn Ranch. They did not mention having stopped there, but before the day was done they arrived at the top of Stull's Falls, the ranch of Walter and Caroline Bremer Stull.[18] They carried Carroll the last three quarters of a mile to the vacant Stull house where Noah stayed with him while Babe walked the three and one-half miles to the Jessie Kyser place. Mrs. Kyser drove him in her Jeep to Allegany where there was a telephone. Babe called the Mills Ambulance Service, asking that they come to the Stull ranch, and he gave Mrs. Carroll the welcome news. By nightfall that Saturday the lost elk hunter, Clayton Carroll, was in a bed at Keizer Hospital in North Bend, surrounded by his family. He had been gone for nine and one-half days, and had been rescued from certain death from starvation and exposure by one of the most audacious rescue attempts that I know of. Harold Noah and Babe Page were able to take quiet satisfaction in their accomplishment. "Without that rubber boat we could not have gotten him out alive," they said.

The rescue of Clayton Carroll brought attention to the Elliott State Forest as a place to hunt and sometimes to get lost. During the

---

[17] The cabin stood back far enough that it could not be seen from the river. Harold Noah no doubt knew about it and was depending on there being food and bedding there. Cle Wilkinson and Baldy Crane completed building it on December 12, 1947. There had been an earlier cabin near there, on the other side of the West Fork but, according to Alice Allen, it had been rendered unusable – she thought a tree had fallen on it. When Cle and Baldy built "Pheasant Cabin" they salvaged the stove pipe and perhaps other items from the older cabin.

[18] The ranch by this time had been purchased from Walter and Carolyn Stull by Tom McCulloch, who a few years later sold it to William Vaughan (Cooston Lumber Company). It remains in the Vaughan family to the present day.

hunting season of 1955 there were two reports noted in the *Coos Bay Times* of lost hunters in the Elliott. The most remarkable of the two was of 17-year-old Donald Miller and Glenn L. Laque who had pitched their camp at Cougar Pass on October 28 (the same place Clayton Carroll had pitched his camp 5 years earlier). They were reported lost a few days later and an extensive search was underway by friends, the Coast Guard, law enforcement, the Coos Forest Protective Association, and loggers in the area. A week later they came to the ranch of George Gault on the West Fork, apparently none the worse for their experience. They had traversed the same distance through the same forest that Clayton Carroll almost died in. The following week the *Coos Bay Times* noted that three other "lost"elk hunters had walked out "OK." They were Robbie Richards, Don Amos, and Joe Blondell who had gone into the Elkhorn Ranch area, shot their elk toward the end of the day, stayed overnight and carried the meat out over the next day, arriving home the evening of the third day. A search party was planned to for the following morning.[19]

One has to wonder what motivated Harold Noah and Babe Page to such extreme measures to save a perfect stranger. Harold Noah was evidently a quite exceptional man among exceptional men. Married and supporting a family from the age of 17, falling the big old-growth Douglas fir with a hand saw all through the 1930's and 40's. Trapping mink during the times when there was no other work during the Great Depression, and after WW-II being among the first to begin using the new gasoline chain saws to fall timber. The old-time fallers were paid by the thousand board feet – by the bushel, as they said. When they started falling with the new gasoline chain saws, they were paid the same price per thousand, but with the new saw they could cut ten times as much, thus making ten times as much money. At the time of the rescue, Harold was falling timber for Herb Laskey's logging company at Lake Tahkenitch.

Harold also had a Caterpillar tractor so that he could log under contract in the summer months, and so he was a "gyppo logger" in its most classic sense.[20] He had his own chainsaw, and his own Cat and he was his own boss most of the time. When the challenge came, as it did in November 1945, and the search was on for the crew of the Army C-46 aircraft, he went far beyond what could be reasonably expected of him and remained on the search for two weeks after it was called off. There was no hope of reward or fame, gold or glory, in doing that. It evidently was what he considered the right thing to do.

[19] See *Coos Bay Times* Nov. 5, 1955, "Young Lost Hunters Are Found Near Allegany," and November 16, 1955, "Three Missing Elk Hunters Walk Out OK."

[20] See my *Sawdust in the Western Woods* (2009) for a full explication of the term "gyppo," its history and current status.

And he did it again, in his successful rescue of Clayton Carroll. But, like so many of the loggers who advanced to the ranks of "gyppo" in those days, he was a Mason. I am not a member of any kind of lodge, and most of what I know about the order I've read in Mark A. Tabbert's *American Freemasons*.[21] According to that book it is basically a self-improvement organization with a mission to "make good men better." During Noah's time, lodge membership was essential for anyone expecting to rise within the management of most of the big lumber companies, and it helped to get the contracts needed to operate as a gyppo, or even to get an hourly job in the woods when jobs were scarce. Masonic brothers were said to "meet on the level, act on the plumb, and part on the square."

What is it that motivates anyone to risk one's life for the benefit of a perfect stranger? The term is alturism, a subject much studied and little understood.[22] There is apparently a biological element in altruistic behavior; it presumably serves some purpose in the evolution of life. And it is enhanced through cultural inheritance, family upbringing, religious traditions, and perhaps lodge memberships. It sometimes saves individuals from certain death – as in the case of the lost hunters on the South Fork in 1871, in the case of the lost flyer Walter Smoyers in 1945, and very certainly in the case of Harold Noah's rescue of Clayton Carroll in 1950. Anyone totally lost and helpless is forced to depend for survival upon the kindness of strangers. If they are lucky enough, someone like Harold Noah may come along and save them.

And how did Clayton Carroll come out of the ordeal? According to his daughter Patricia, the doctors saved his leg but had to amputate several toes. Immersion foot is usually caused from wearing tight-fitting and non-breathing boots over lengthy periods. The rubber boots that Carroll was wearing were the perfect agent. At first the feet become numb and turn red or blue, then they begin to swell until the boots cannot be removed. Blisters and open sores come next, and if the boots are not removed and the feet allowed to dry, gangrene is next and if that is not treated – usually by amputation – death ensues. Immersion foot is very serious indeed.[23]

Clayton's daughter told me that her father worked only at odd jobs after his ordeal, going into a semi-retirement. He was, after all, 58 years old and nearing retirement age in any case. He had to wear special shoes, and he had

---

[21] Mark A. Tabbert, *American Freemasons: Three Centuries of Building Communities*, National Heritage Museum and New York University Press, 2005, pp. 9, 164-5.

[22] See "Is Goodness in Your Genes?" by H. Alen Orr (*New York Review of Books* LVII, 15 pp. 32-34).

[23] See *Military Dermatology*, Chapter 4, "Immersion Foot Syndrome." Office of the Surgeon General, 1994.

poor balance. Walking was painful and he used crutches for a time, then a cane. He never hunted again, but found great pleasure in fishing. I asked if her father maintained an acquaintance with her rescuers and she said, "Yes. They remained close friends for life." In fact, she said, her parents bought a house in Libby next door to the house of Babe Page. Clayton Carroll died on September 6, 1961, at the age of 68. He had been given an additional eleven years of life with his family, which Harold Noah and Babe Page probably considered ample reward for their efforts.

Harold Noah and Babe Page the day after their daring rescue of Clayton Carroll
CCHMM 955-1.11833

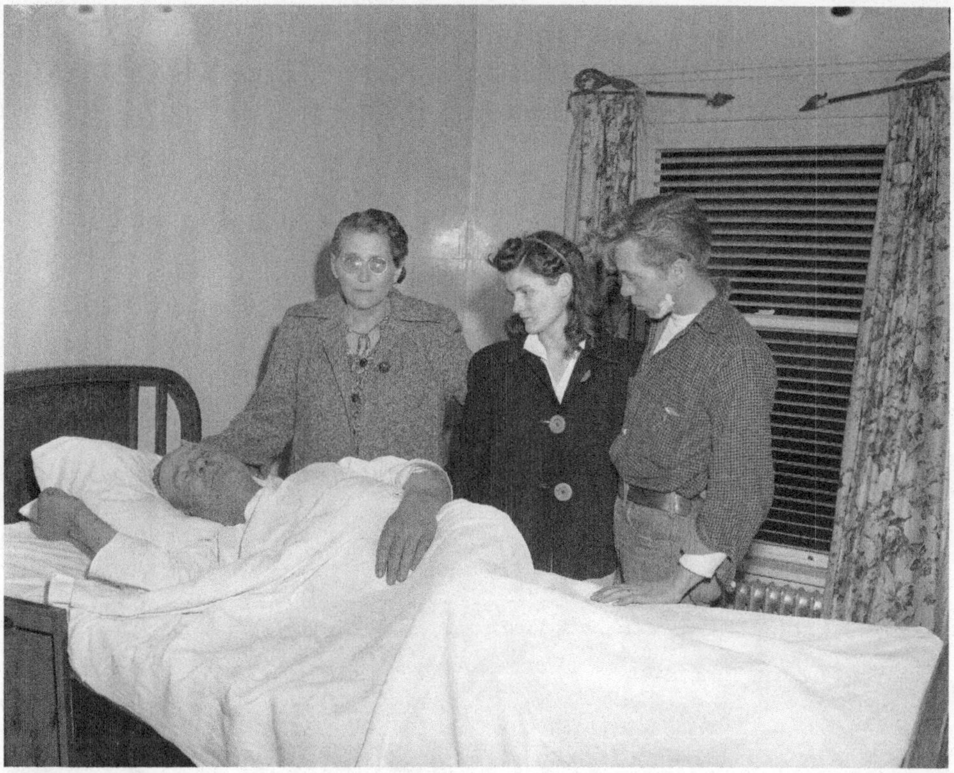

Clayton Carroll at the Keiser Hospital in North Bend following his rescue. Standing at his bedside is his wife Hester and daughter Patricia with her husband James Wallin. CCHMM 955-1.11822.1

Map VII,
Aircraft Crash Site
(about one mile west of Trail Butte in Section 7, T 24S R 11 WWM)

Courtesy Jerry Phillips

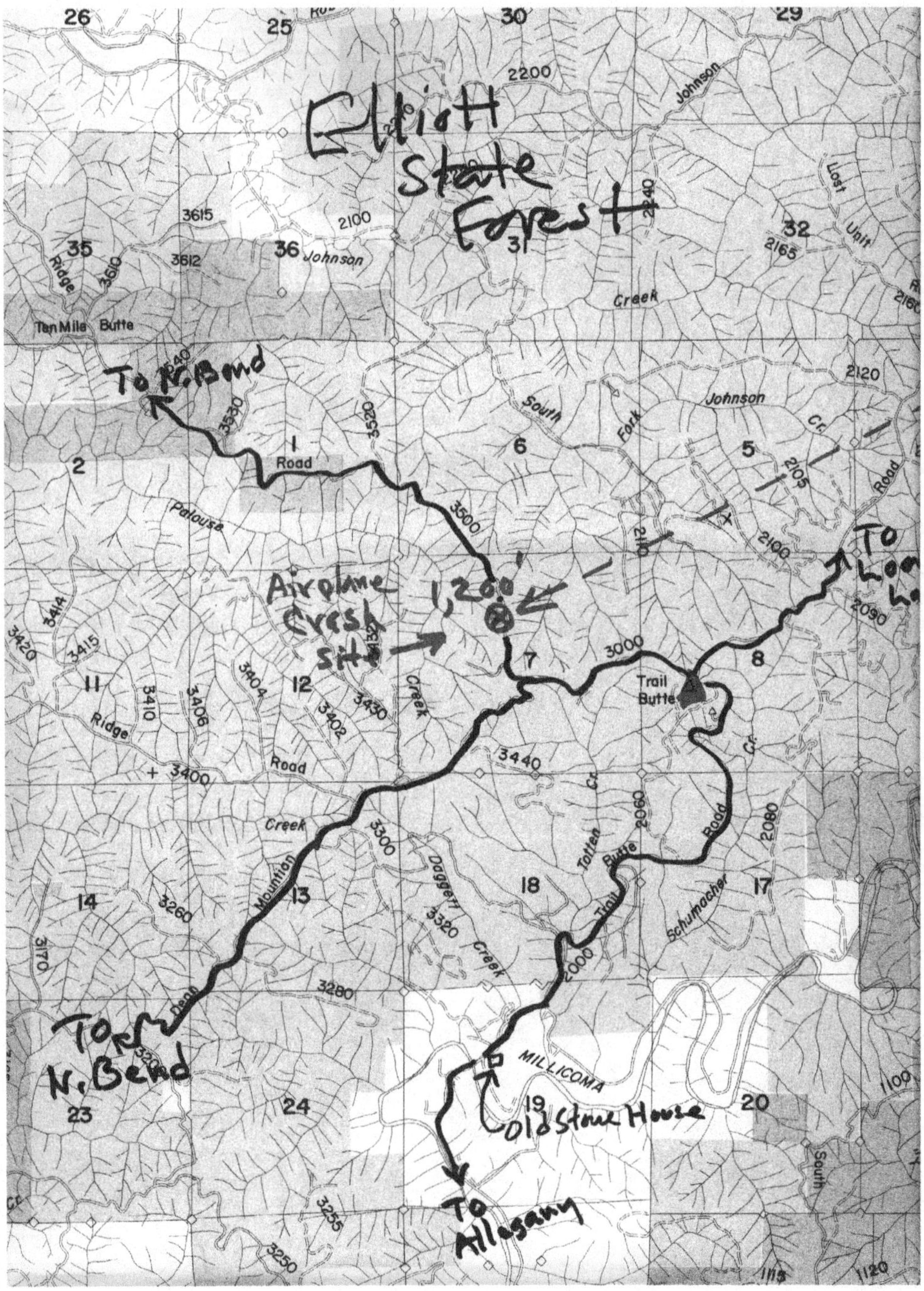

Chapter 6

Lost for 23 Years

A small aircraft disappears in 1954 and is rediscovered in 1977

James Frank is a name almost too comonplace to be real, but James Frank was not commonplace. He did quite a bit in his very short life. Born September 12, 1934, at Scottsbluff, Nebraska, the son of Alex and Lorraine Frank, he graduated from Marshfield High School in 1952. In the yearbook each of the seniors traditionally left a "will" to the next class, in the form of a witty one-liner. Frank was evidently anxious to get on with his life because in his "will" he stated, "I, Jim Frank, will leave, willingly."[1] After graduation he wasted no time. He got a job clerking at a grocery store and four months later, in October, he married 19-year-old Dorothy Ann Durflinger, who had graduated from North Bend High School the year before. Their wedding was at the Presbyterian church and Dorothy's older brother Bill and his wife Jean were the witnesses.

Dorothy was working in an office and it wasn't long before their first baby, Rodney, was born and not long after that she was pregnant with her second. Fast, young marriages right out of high school were all too common in those days before the pill, and some of them worked out and some of them didn't. Jim landed a job with the Coos Bay Post Office, and the young family moved to 447 Bruell in Empire. From the evidence, it would seem that his mind was on things other than family when he began taking flying lessons at North Bend Airport and went into partners with a co-worker, Jim Long, on buying a 1947 Ercoupe, advertised as the safest airplane in the world. What Dorothy thought of this is not on record.

Civil aviation was being highly promoted during the decade following WW II, and Jim was eager to be a part of it. The Ercoupe was a small two-seater, a low-wing monoplane with twin tails. It had an aluminum fuselage and fabric covered wings and the one that James and his co-worker bought was silver with red trim on the ailerons. It was a very neat and attractive little airplane designed specifically for the casual flyer. Flying it was supposed to be as natural as driving a car and it was almost impossible to stall or spin. The earlier models didn't even have rudder pedals – the aelerons and rudder were interconnected so that the plane could be steered like a car, without the use of the pedals.

[1] 1952 *Mahiscan*, Marshfield High School Yearbook.

James Frank, 1934-1954

Frank received his pilot's certificate the first week of May, 1954. Three weeks later, on 25 May, he filed a flight plan from North Bend to Eugene, intending to fly on from there to Beaverton to visit his father. This would be his first cross-country solo. Wearing a leather jacket and sneakers, he took off from North Bend at 1:10 p.m., but there was a heavy cloud cover with intermittant showers through the day. Twenty minutes after take off he was under a 500-foot ceiling on the Umpqua River, trying to find his way back. He was last seen about 200 feet over the bridge at Scottsburg and was not seen again for twenty-three years, lost under the thick foliage of the Elliott State Forest.

It was too late to begin the search that first afternoon but the following morning, Wednesday, six private airplanes, a Coast Guard search party, and a number of private search parties began looking, primarily between Dean Mountain and the Umpqua. The Oregon Air Search and Rescue Group under its state director, Earl Snyder, took charge with headquarters initially at the North Bend Airport where Frank's wife waited with their 14-month-old son, doing her best to show faith that he would return safely.[2]

There was only one credible reported sighting of the plane after take off. Bob Thomas, a pilot, said that he saw an Ercoupe flying up the Umpqua River about 200 feet over the Scottsburg bridge sometime between 1:30 and 2 pm Tuesday. He said that there was a 500-foot ceiling there at that time. By Thursday there were at least 25 airplanes from the Civil Air Patrol and Oregon Air Search Rescue Group, including ten from Springfield. The search began to concentrate on the area between Elkton and Drain.

As is usual in cases of missing aircraft, much of the search effort was devoted to running down false sightings. Loggers reported an aircraft seen over the North Fork of the Coquille River so search parties headed into that area. Something metalic was sighted from the air about six miles east of Scottsburg by Douglas County District Attorney John Pickett. He landed, and went 10 miles through the brush and was 50 feet from it before he realized it was a large tinfoil

[2] My account of the search is drawn from the *Coos Bay Times* newspaper articles of May 26 through June 3, 1954, and the weekly *Empire Builder* of May 27 and June 3, 1954.

cross used by foresters. Another bright metal object was seen on Smith River, one mile east of Reedsport. The search was very widely scattered, as is common in searches for downed airplanes.

As the days passed, the search effort followed up on one false lead after another. One week into the search, headquarters moved to the Douglas County Forest Protective Association at Elkton. On June 2, newspapers reported that 40 men were "pushing toward the missing plane" east of Elkton where Roy Cole, a logger, remembered hearing a plane's engine "sputter and die" a week earlier.

Ercoupe 415C, "Standard" two-place model

Finally, on June 3, nine days after the plane disappeared, the search was called off. Douglas County Sheriff Lee Shipley said "the area was too vast to get anywhere near a complete coverage." It was late spring and the foliage was too thick and the area was too large. Other than random aerial observation, there had apparently been little attention to the Coos County part of the Elliott State Forest.

The Elliott State Forest had come into being in 1929, the culmination of efforts of State Forester Francis Elliott, whose dream was to have the burn of the old Coos Bay fire of 1868 managed as a commercial forest for support of the state's schools. His idea resulted in the first state forest in Oregon and the first in the nation that was based primarily on management of school lands set aside under provisions of the Land Ordnance of 1785. It was a profoundly radical idea that has proven itself over time.

The Elliott State Forest comprises about 92,000 acres of timberland. Most of it had burned in the great Coos Bay fire of 1868 and by 1954 much of it had grown back to solid stands of 65-year-old Douglas fir. It was all natural re-growth and therefore was not as perfectly distributed as industry foresters would like it, but it was very good for a gift of nature. The network of trails established after the fire by pioneer George Gould had grown over, and some of them were lost.

Commercial logging didn't get started in the Elliott until 1955, and roads had not extended much beyond the fire access roads the Civilian Conservation Corp (CCC) had built during the late 1930's. It wasn't until after the great Columbus Day windstorm of October 12, 1962, that major roadbuilding and accelerated logging began. During that storm, winds broke all records – the anemometer at the Air Force radar station at Hauser hit 152 mph, the top of its scale. No one knows what the peak winds were that day, but they blew down

11 billion board feet of timber in western Washington and Oregon. It was estimated that 100 million board feet of it was blown down in the Elliott State Forest, in a random pattern that affected most of the western half where there had been almost no logging roads built up to that time. The urgency of building roads and getting in to salvage the downed timber was due to the perceived danger of an infestation of the Douglas fir bark beetle. Foresters were concerned that the beetle would get into the fallen trees and then spread into the green timber. That fear was not realized, at least partly due to the salvage logging, which was completed in about four years.

By the time of the Columbus Day Storm in 1962, James Frank had been missing for eight years, and was legally dead. His widow remarried a few years later. The major road building program for the salvage effort was under way beginning in 1963.[3] Among those new logging access roads was the Palouse Ridge Road – the 3500 line – that runs from near the Trail Butte Lookout down the ridge dividing Palouse Creek (which drains into Haines Inlet of Coos Bay) and Johnson Creek (which drains into Tenmile Lake at Templeton).[4] With all of the logging and road building activity, no one ever saw a hint of an airplane on the ground and no one was looking for one. James Frank and his little Ercoupe had been long forgotten.

After the Palouse Ridge Road was built, another ten years went by with timber sales here and there along the road, from time to time, with the objective of responsible management of the resources. Finally, on March 23, 1977, a routine timber sale was being laid east of the road, about a mile from Trail Butte Lookout.[5] Its boundary was running down a gully about 200 feet below the east side of the road when, to the surprise of the forestry workers laying out the sale boundary, it ran right over the top of what remained of the aluminum fuselage of the little Ercoupe. "If they hadn't walked within 30 feet of it, they'd never have seen it," Jerry Phillips was quoted as saying at the time.[6] According to Vince Kohler, staff

[3] See Jerry Phillips, *Caulk Shoes and Cheese Sandwiches*, p. 310. By 1972 some 450 miles of roads had been built.

[4] The road numbering system used in most industrial timberlands uses 1000, 2000, 3000, and so forth for the trunk roads. In this case, the trunk road leaving the county road on the West Fork of the Millicoma and running up to Trail Butte, using the old trail to Elkhorn Ranch, is the 2000 line. At the ridge at Trail Butte, the 3000 line runs to the west and down Larson Inlet toward Coos Bay. The 3500 runs northwest along the ridge as described above. Most of the roads also have names, which do not necessarily coincide with the numbering system. The logical name for the 3500 line is the Palouse Ridge Road.

[5] In the NW 1/4 of section 7, T 24S, R 12 WWM.

[6] Coos Bay *World*, March 24, 1977, p. 1. See also Jerry Phillips, op. cit., p. 396.

writer for *The World* newspaper, it "was crumpled like so much pale green and silver aluminum foil. Still visible on the tail was its identification number: N-3850-H." He continued, "The plane's wheel assemblies stuck straight up, like the stiff legs of a dead animal. The rubber wheels were whole. They still turned."

The next day a caravan of official vehicles including the sheriff's Jeep, a state police car, a forestry truck, and a Coos County Emergency Services van, drove up the winding state forest road to the site of the crash. Vince Kohler said they began by prying the plane apart with shovels and axes, and they passed Frank's bones and personal effects to the State police trooper who carefully placed them in individual plastic bags. The bones, Kohler said, were "dry, brown, and wood-like." They found his wallet and his pilot's certificate, and the ignition keys to the plane. Frank had been wearing his leather jacket and sneakers. When notified of the find, his widow Dorothy said, "It's good to know what happened after all this time. It sort of closes the book." Indeed it does.

The plane had hit the top of the trees and from the visible scars it appeared to have come in from the northeast on a straight course toward the North Bend Airport from Elkton at an altitude of 1000 feet. Palouse Ridge at that point is 1200 feet high, and about eleven air miles from the airport. James Frank was 19 years, 8 months, and 13 days old when he died and it was 23 years later when they removed his bones from the cockpit of the Ercoupe, the safest airplane in the world.[7] Vince Kohler provided an epitaph: "Lost in the springtime of the year 1954; in the springtime of his life."

[7] *The World* newspaper consistently gave his age as 24 years but in the application for their marriage license Frank claimed his age as 18, and his date of birth as Sept. 12, 1934.

PART TWO: After the Roads

Part Two consists of Chapters 7 through 12, all of which are set after the Columbus Day Storm of 1962. This set of stories is characterized by the existence of hundreds of miles of logging roads within the Elliott State Forest and the Weyerhaeuser Millicoma Tree Farm. It is also characterized in all but one of the incidents as being vehicle related. And in all but one of them the object of the searches are female, two of whom were small girls.

Map VIII

Glenn Creek Junction

USGS map, Ivers Peak, Oreg. 1955

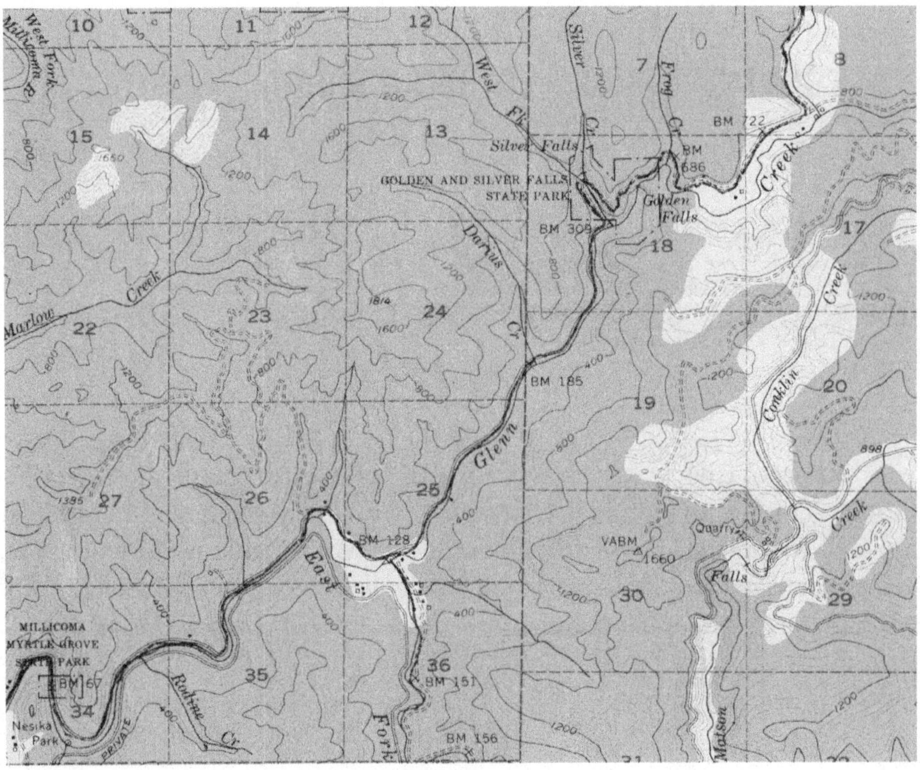

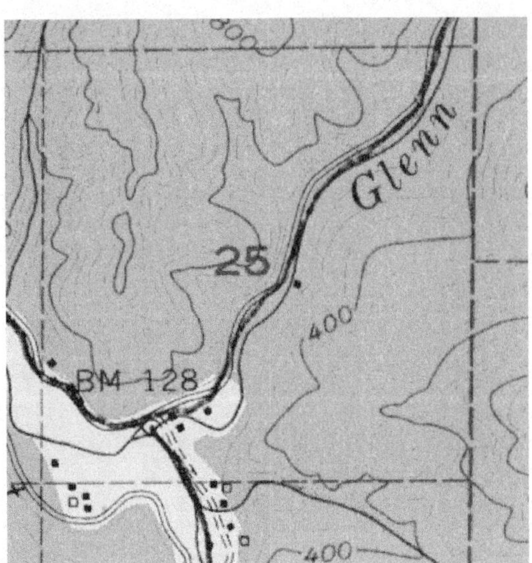

Section 25, Township 24S Range 11W.W.M.

The county road runs northwest on the west side of Glenn Creek. The house is on the east side of the creek (indicated by the dot opposite the number 25). Ondine, completely turned around, wandered in an apparently zig-zag pattern, tending always in a southwesterly direction until, three days later, she came out on the East Fork of the Millicoma directly across the river from the home of Wilma and Fred Lund.

scale: each section is one mile square.

CHAPTER 7

ONDINE: Christmas, 1977

A seven-year-old girl lost on Glenn Creek for three days and nights in December

Ondine was lost and presumed dead when she came out of the woods across the river from the house of Wilma Lund. Wilma was my wife's best friend and she died April 7, 2010, at the age of 95. I wanted to give a little talk at her memorial at the Allegany potluck, and tell among other things about the rescue of the little girl. I looked up the newspaper articles about it and found her name, then looked it up on the internet and found her. She is pretty high up in management at the UCLA Medical Center. I e-mailed her and received a wonderful exchange of communication, which I used at Wilma's memorial and I am using as the core of this article. I was emotionally pretty much involved at the time of the search for her and at that time I wrote a few scribbled pages which I hadn't seen for 33 years. I searched for those note pages all over the house for several days -- knowing that I hadn't thrown them out. I finally found them paper-clipped with a bunch of unrelated newspaper clippings in the very back of a seldom used filing cabinet, and with those notes and the e-mails, I was able to piece together the following story.

The girl and her mother were visiting. It was Christmas Eve and they were with their friends who lived way up on the other side of Glenn Creek. There is no road to the other side; there were no electric lines, and there was no telephone. The people who lived there over the years lived there because that is what they wanted and they never changed it. The big storm that had recently passed deposited eight inches of rain on the Coast Range. The springs and creeks and rivers were all up. Glenn Creek was running high and muddy, three feet above its usual for this time of year. The rain forest was green and wet and the sound of turbulant waters dominated. A high pressure system between two storms rendered the day clear and fairly warm – warm for late December. The sun shone and Christmas was tomorrow, which in 1977 was the shortest day of that year.

The radio was full of news of a fishing boat that had broken in two off the Southern Oregon Coast. It was the *Holy Redeemer* – a name that resonated with the season and probably for that reason it was very much in the news. The Coast Guard had called off its search for survivors. Far offshore in the Pacific another storm was coming, as was normal for the season.

Back on Glenn Creek there were presents. Some were for the little girl, who was seven, and some were for David, the ten-year-old boy who lived there. These were the only children on that Christmas at that place. They tried not to show their excitement too much, but it was hard. At the house it was easier than ever to get into trouble and they had been cooped up for too long during the recent storm. Ellen, the girl's mother, and Renée, the boy's stepmother, were talking with two other ladies who were visiting. They had made cookies and they sent the kids outside to play while they made the frosting. The girl had on her checkered play trousers and a very light yellow cotton sweatshirt. She slipped her bare feet into a pair of well fitting rubber boots and ran outside.

Outside there was lots of room. There were the woods above the house that went on forever – the Mahaffy timberland, logged during the 1930's and almost ready to be logged again, adjoining 215,000 acres of Weyerhauser timberland and more, miles and miles more. David knew the woods around the house because he lived there and played there all the time. At least he knew some of the woods. He had never been into the deeper part of them, and he didn't know what was there and he had really not given it much thought. He only knew his part of the woods.

People had lived there off and on for many years. It had first been settled 80 years earlier when Charles Crane was looking for some land off by itself. The places further downstream were pretty well taken. Charles had to go beyond the others, on up through the brush, until he found

Charles Crane

inside the steep canyon of Glenn Creek a few flat acres. There he staked his claim. It was the first place above the homestead of old W. W. R. Glenn himself, who was first, and who gave his name to the creek. Next above Crane was John Hendrickson and next above him was George Schapers whose bachelor cabin was at the present site of the parking lot at the Golden and Silver Falls State Park. There were others above the falls, but that is another story.

Charles Crane was a frontiersman who had his one moment of fame in 1918 when his picture appeared in a nation-wide advertisement for the Boston Woven Hose and Rubber Company's "Good Luck Rubbers," gasket rings used in home canning. He had sent a jar of bear meat canned by use of the rubbers to the company in Boston with the following letter, "gentlemen: I have some canned Bear meat, Venison, and other game also some vegetables and fruit left from last years canning. I am pleased to say I have not lost a single can on which I used your good Luck rubbers. I am Yours Sincerely Chas. Crane." The advertisement had the heading, "The Story of Mr. Crane, the Hunter" and it stated that "the bear meat was fresh and delicious after its journey across the continent."

All the settlers on Glenn Creek were frontiersmen and they all ate bear meat when they could get it, but it was the timberland that made the area economically significant. The timberland on Glenn Creek that most closely surrounded the Crane ranch belonged to the Waterford Lumber Company of St. Paul, Minnesota. During the early 1930's they built a sawmill on Coos Bay, called the Martinez Mill, located just north of the Central Dock. The logging was done by Brady and Neal Logging Company, who in 1935 built a fairly large logging camp at Glenn Junction (the informal name given to the mouth of Glenn Creek, where it enters the East Fork of the Millicoma River). The Brady and Neal camp operated until the Martinez Mill burned in the early summer of 1941. I was seven years old at that time and my family moved into the Brady and Neal camp immediately after it was closed. We lived in Roy Neal's house for a year and I often played in the abandoned Glenn Creek schoolhouse, which was still quite intact even though at that time it had been closed for 16 years. All of the Crane children had attended that school and Lossy, the youngest girl, was the last student there when it was closed in 1925.[1]

The timber from Glenn Creek was logged directly into the creek and by means of a splash dam the logs were driven down the creek into the impounded waters of the Lockhart splash dam on the East Fork of the Millicoma River. When that dam was periodically released, the artificial freshet it created drove the logs ten miles further down the river to Allegany and tide water where they were rafted and towed to the Martinez Mill. Brady and Neal also used logging trucks, the first to use them on the East Fork. They built a road up past the Glenn Creek School to tap the timber on the steep hillside above the schoolhouse, hauling the logs down to the camp where they were dumped into the impounded water from the Lockhart dam. Traces of some of the old spur roads were still descernable as late as 1977, totally grown up to brush and new timber.

After logging, it was usual in those days for the owners of the timberland to stop paying taxes on the now useless property, which would in due course revert to the county. Ben Mahaffy, an astute rancher from further downriver, sometimes bid on those useless logged off timberlands, and over the years he accumulated a fairly large

---

[1] Charles Crane named some of his children after the cities they were born in. Thus, Baltimore Calvert Crane (Baldy) was born on Calvert Street in Baltimore. Los Angeles (Lossy) was born in Los Angeles.

holding of potentially very valuable forest. In 1977, it was the Mahaffy Tree Farms that owned the former Waterford timerland that was above the house, and the new timber that came up was now 40 years old and almost ready for harvest.

Over the years the Crane property was alternately vacant and rented and leased and squatted upon. Among the tenants at one time was John and Emma Noah – parents of Harold Noah, the hero of the rescue of the elk hunter Clayton Carrol (Chapter 5). Harold probably lived there during his youth (I'm not certain of this, but he was emninently familiar with the area). The place was eventually bought and sold until it fell into the hands of a speculator who waited for the value to rise. And eventually the value rose. This vacant land on the other side of the creek, without a road or a bridge or a power line became valuable because a new generation was yearning for wilderness of its own, before it was all gone. Finally, in about 1976, a young counter-cultural woman bought it for the same reason that Charles Crane had settled on it 80 years earlier. It was the only place left that wasn't taken. And it was private, and it was isolated and it was just what she wanted.

And so on this Christmas Eve of 1977, the boy and the girl played. They played hard because they had to spend their energy while they waited for Christmas, the unbearable wait for tomorrow and the presents. They ran. They ran up the hill and through the maple and myrtle and into the sword fern and they ran down the hill and along the hill and they ran and they ran. David was faster. He was a little older. And he ran and he turned and he twisted in his course, down, around, toward the turbulent waters of Glenn Creek. Then he stopped. He was near the creek, downstream from the house a long ways. He was alone. It had been a wonderful chase and he had won. He had got away from the girl who was chasing him. He waited awhile for the girl to show up. He listened, but all he could hear was sound from the rushing and swollen creek. He started back toward the house. Then he stopped and listened some more. Strange, he thought. She should be coming out by now, but maybe she went back to the house. He had been gone not more than fifteen minutes.

A neighbor who lived part-time near the mouth of Glenn Creek was driving up the road. He was going to help old Henry Crump save a woodlog that had come down in the freshet. Henry made most of his living cutting firewood and he was always on the lookout for a good woodlog. The neighbor didn't know where the log was, and thought he might have passed it, so he turned around at the wide place in the road across the creek from where the boy had come out of the woods. The neighbor was thinking about a coincidence announced earlier on the radio news. The wife of the skipper of the *Holy Redeemer* had been awakened by the telephone. As she answered it, she discovered the house was on fire. She saved herself and the chidren before the house was destroyed in the flames. It was the Coast Guard that had called to tell her that her husband was lost at sea and there was no hope. The search for survivors had been called off.

As the neighbor turned around his pickup, he also thought of relatives he would visit in town this Christmas Eve. He was in a hurry. He wanted to help old Henry save the woodlog, but he was also in a hurry to get to town. It was only a little after two o'clock but the sun was already behind the trees on the ridge. There was a real chill in the shaded canyon. As he shifted into reverse the second time, he heard – or thought he heard – someone yell above the roar of the creek. He backed up four feet and shifted into low. He heard it again. Must be the boy across the creek, playing, he thought. As he pulled ahead he heard it again. What's going on over there, he wondered. But he was in a hurry to help save a woodlog, so he drove off down the road.

A half an hour later the menfolk of the Renée's place across the creek returned from town amid laughter and gaiety befitting the season. They were clowning and taking pictures in the road. The womenfolk came to the creekbank on the other side and announced over the roar of the creek that the little girl had been gone about 45 minutes and no one knew where she was. This unwelcome news brought no immediate action, but the women were concerned and soon the sight and sound of the rushing waters triggered an image of tragedy even in the minds of the reveling men. Instantly sober, a search was on.

A couple of hours later a discouraged and worried Frank Stewart knocked on my door. We lived five miles downstream and Frank often used our telephone. He told me that a seven-year-old girl who was visiting at their place was lost, had disappeared completely. He told me that the girl and her mother lived in Charleston and the girl's name was Ondine. Frank was unsure whether he should call the Sheriff so soon, but after talking it over we both agreed that he should call and try to get help. He called the Sheriff's office and the Coast Guard.

I went back with him, but I didn't go across the creek to the house. During summer months they used a low-water crossing at the south end of their property, but when the water was up they used a raft tied to a tree to float across nearer the house. We went along the creek for the mile between their house and Glenn Junction, assuming that if the girl was all right, she would naturally come down to the creek. Although she couldn't get across it, we assumed that she would be someplace on the other side. There were many pools and rapids that were quite ominous, however, and the thought that she may have fallen in and drowned was never far from our minds. At one point Frank became impatient and demanded, "Are we looking for a cadaver?" I said I didn't know. It just seemed that the creek was the place to look. Within an hour it was too dark to see what we were doing and I returned home.

About 9 o'clock that evening I heard a noise outside. It was a Deputy Sheriff looking for directions to the place where the girl was lost. He had run off my driveway, hopelessly stuck and we had to call a wrecker to get him out. My driveway was badly torn up as a result and that was all I knew of help from the Sheriff's search and rescue that evening. Temperature recorded that night at the North Bend airport, with its moderating influence of the tidal waters of the bay, was 47 degrees. The temperature at Glenn Creek was probably about 7 degrees lower. Actually, the Coast Guard helicopter was out fairly soon and was there for several hours that night. Search and Rescue volunteers came out with dogs, but they kept going down to the river. They stayed until late that Christmas Eve while Renée and Ellen and the other ladies made coffee and tamales to keep them going.

The next morning – Christmas – was the beginning of another clear and comparatively warm day. At the airport it got up to 60 degrees before it started down again. The boy tore open his presents with a studied lack of interest in them. Ondine's presents lay where they were, under the tree, and it was a very somber Christmas. Around 8 o'clock a string of cars and pickups went past our house. It was the Sheriff's volunteer search and rescue team, with entire families. I didn't go up to help because there were at least 25 of the volunteers who were presumably trained in search and rescue and had tracking dogs. They would find her. Frank told me later that they were exceptionally dedicated and did everything in their power to find the girl. The dogs would follow her scent as far as the large sword fern and huckleberry brush – about the last place that the boy had seen her – then would turn back toward Glenn Creek. For that reason the search tended to concentrate along the creek and the immediately adjacent hillside, although they went far up into the hill, back and forth, but found no clues. They apparently did not go down the east side of Glenn Creek very far, or they would have come across her. She was there.

Coast Guard helicopters were overhead during much of the time, looking down on the top of a solid mass of trees and brush. Much of it was a triple-layer canopy consisting first of 120- foot high, 40-year-old Douglas fir and hemlock – the natural regrowth from the logging done by Brady and Neal during the 1930's. Below the crowns of those conifers was a mix of broadleaf trees. At the bottom, the layer that Ondine would be under, were solid hedges of black huckleberry, sword ferns twice as high as her head, and other species of brush. At seven years old she was only about 3 1/2 feet high and weighed about 45 pounds. She was a mere speck on the floor of an impenetrable jungle.

The primary area of the search consisted of little more than one mile to the south of the house, along the east side of Glenn Creek. It could extend no more than a half mile up from the creek, a fairly steep hill that averaged about 30%

grade until it met the almost vertical rock outcrops near the top of the ridge. It was not a very large area, as search areas go, but none of the ground was visible from the air. If she was down there, she was beneath that canopy, under those huckleberry bushes and among those giant sword fern, completely disoriented and with no point of reference. She might have been found by someone with a machete going through on the ground, calling for her at regular intervals, but no such attempt was made. At the end of that Christmas day, in consultation with the girl's mother, the search was called off by Sheriff Les Miller. It had been barely 30 hours after she was discovered missing with probably not more than 12 hours of actual search.

Hypothermia occurs when more body heat is being lost than the body can generate. It is said that the average person standing naked in still, dry, 50 degree air, can maintain normal body heat of 98.6 degrees. For every degree the air falls below 50, the body temperature drops a corresponding degree. Once the body is below 95 degrees, hypothermia begins and if the air temperature continues to fall, death will become inevitable. Ondine was dressed very lightly, her upper body covered by a mere cotton sweatshirt and no socks on her feet. She had been exposed to air temperatures below 50 degrees for much of the 30 hours, and during that first night the temperatures were around 40. The Sheriff's Search and Rescue experts assured the mother that there was very little chance of survival if the victim is not found within the first 24 to 36 hours. They were wrong there. But, they added, "a lot of people end up walking out about 3 days later." They were right right on that one!

That night, the second night that she was missing, it was much colder. The airport recorded 37 degrees, and given a good 7 degrees lower on Glenn Creek, it was a couple degrees below freezing there. It was the night of the full moon, but it would be almost midnight before that full moon could have risen above the 1200 foot ridge that separates the Glenn Creek canyon from Matson Creek to the east. In that canyon the sun and the moon come up late, and go down early. It is usually chilly, damp, and gloomy in winter, day and night, but the next day was bright and clear though much cooler than the previous day. At the airport a high of 52 degrees was recorded, but in the gloomy Glenn Creek canyon, peak temperatures only last a very short time during mid-day before the chill and shade sets in again. It was Monday, and a holiday because Christmas had been on Sunday that year. Everybody stayed home with their families, and nobody was looking for Ondine.

When that day ended, the third night began. This time the airport temperature was recorded at 39 degrees, which would put the Glenn Creek temperature at around freezing. There was a small-craft advisory and before daylight the next morning the beginning of a new storm front was showing up, raining a slow, cold mist. It didn't look good for someone dressed as Ondine was dressed, and no one had been looking for her at any time during the past 36 hours.

I got up that morning and looked out at the change of weather, a very cold rain, and I was depressed. The rescue chief at another search for a child several years later was quoted as saying, "There is something about a child being lost that just hits at the heart of people."[2] I can certainly vouch for that sentiment. If the girl had survived up until now, it would have been because it had not rained and she was essentially dry. Water, it is said, conducts heat 240 times as fast as dry air. It would seem impossible that she could make it through a day of this very cold rain after the exposure she had already experienced.

I was at work at my job in North Bend for about an hour that morning when my wife called me. She said that Wilma Lund, who lived on the East Fork across the river from the location of the old Glenn Creek schoolhouse, had found the girl and she was alive and at the hospital. This whole event had shaken me but I had resigned myself to the worst. In addition to drowning, or hypothermia, there are bears and there are cougars, and the imagination knows no bounds. My wife's call stunned me with hope. I had to see the girl and so I rushed straight to the hospital. A doctor was just leaving her room as I arrived and he was shaking his head as though in disbelief. I opened the door and her mother was sitting there. I asked how she was. Her mother said, in a tone of wonder, "She's going to be all right." I walked to her bed and looked down at her and was surprised at the good color in her face. She looked back with bright intelligent eyes and I could only say, "We're glad you're back." I never got an accurate story of how she survived until thirty-three years later. I found her e-mail address on the internet and contacted her to say that her benefactor, Wilma Lund, had passed away at the age of 95 and that I would like to tell the story of the rescue at a commemorative potluck the community was giving. Ondine told her story as follows:

---

[2] Del Sparks, assistant rescue chief on the Nathan Madsen search quoted in the *Eugene Register Guard* October 27,1989.

ONDINE'S STORY

by

Ondine Eaton

(from e-mails to Lionel Youst 4-28, 4-30, and 5-2-2010)

Ondine Eaton in 2010

Hello Lionel,

Thank you for contacting me, it's been such a long time and I was so young, that I had forgotten the names of the people that helped me that day. I'm sorry to hear of Wilma's passing, she was very kind to me and I never forgot the day that I finally emerged from the woods and saw an actual person.

I'd be happy to share any information about the event that you want to know. I am alive and well (obviously) and working as a project manager at the UCLA Medical Center in Los Angeles. I am now 39 years old and have had only one lasting effect from the days I spent in the woods and that is a little nerve damage in my big toe that I don't even notice. When I arrived at the hospital I had frostbite in both of my feet and could not walk for several days. There was frozen mud inside my rubber boots and I had no socks on as we didn't plan to be out for long that day. They had to cut the boots off of my feet and bathe them in warm water every few hours for the first couple of days. It was extremely painful. I was very lucky to have kept my feet. Little kids are tough and heal quickly thank goodness.

Most of the newspaper articles from that time are really inaccurate by the way. I don't how they got details so messed up. I was with my mother's friend's son, David, and we just went for a short walk before our meal was ready back at the house. I believe we had been shooed out of the house by our mothers while they cooked and talked. David was only [10] years old and didn't realize how small I was in comparison to the ferns in the forest. We were running around on the trails and chasing each other when I must have veered off and gone the wrong way. David looked for me but couldn't see me anywhere, so he went back to the house thinking I had gone back. Because I was so short and the brush was so tall near the house I couldn't see which way was back to the house. So I kept walking....and walking. By the time everyone realized I was lost, I had gone a long way and couldn't hear them yelling for me.

The next few days were very cold and very frightening as you might imagine. To make matters worse, on several occasions I saw the search helicopters and screamed and waved my arms, but they never saw me because I was so little and the brush was so tall. I also heard dogs barking on the second day and searched for hours for where the sound was coming from, leading me further away from home.

My time in the woods was pretty much the same the whole time. The first day, when I realized that I was well and truly lost, I panicked and cried for about an hour or so, then got over it and decided I had to keep looking or freeze to death. After that it was just a matter of keeping moving during the day (in a random zig zag pattern apparently, so much for sense of direction) to try and find my way back and not freeze at night. At night I pulled my arms inside my little shirt and wrapped them around my knees, then put my face inside the neck of my shirt to catch the heat from my breath and keep my face warm. I would pass out from exhaustion now and again during the night and get a little sleep, then get up at dawn and start moving again. I drank from the river (carefully, it was flooding) or the small streams when I got really thirsty and found a few wild chanterelle mushrooms to eat occasionally. Other than that, just walking and calling out constantly, hoping someone would hear me. By the end of the second day, the fear and hunger were mostly gone, replaced by one single goal...to keep moving and find something before I died from the cold. Years of living in a fishing village had taught me that people can and do die from hypothermia all the time and even at that young age, I knew it was way too cold to survive long.

The only thing that happened the morning before coming to Wilma's place was that it was beginning to rain. I don't think anything else was different, lots of walking at dawn and being wet and cold. As far as the birds, I told the news reporters that the birds would wake me up every morning very early and that was the only Christmas present I got while out in the woods.

The day that Wilma saw me, I was walking along the river's edge and saw some of the pipes or something that ran down to the river from their house and went to look. I saw someone standing on the opposite side of the river and yelled for help. That was Wilma and the man I assumed was her husband. They took me to their house and called the police, then gave me some soup and a blanket. Next thing I know, there's a helicopter and I'm being stuffed into it for an air lift to the hospital. I remember being very annoyed that they wouldn't let me look out the window, I had never been in a helicopter before!

The hours after that at the hospital were spent being mummified in blankets and poked by doctors, while my mom completely freaked out. They didn't let me move after that and I was confined to a bed for what seemed like eternity. Some news crews came and asked a lot of questions. I did manage to annoy the nurses by pushing the call button all the time out of boredom, but that was about it for excitement at the hospital.

My mother passed away 3 years ago from cancer, but she was always very grateful that I had made it out of the woods alive and I know that she never forgot how lucky I was to have run into Wilma. Please give my regards to Wilma's family and friends at the potluck and let them know that I am thinking of her.

Sincerely,
Ondine

Aftermath

How did she do it, three days and three nights at the end of December, totally lost and invisible, under a triple canopy of trees and brush that a grown man could not have gotten through without a machete or a chain saw? For one thing she was small enough to walk upright under a wall of brush that a grown man could not have gone through. She was lucky that it was no colder than it was and that the rain didn't come until that last few hours before she had found her own way out of the tangle. It helped that she was very bright and could figure things out, and that she had an iron determination to survive.

People who are totally disoriented have two choices as to strategy. They may stay put, and wait to be found, or they may start off walking. Either strategy may succeed, or it may fail. Back in January, 1917, above Golden Falls – about three miles above the place on Glenn Creek where Ondine was lost – a teacher and her 7th grade student were lost overnight on Frog Creek. Dora Brown, the teacher at the Golden Falls School on the Tyberg place and her student Agnes Wilkinson had been out for a Sunday walk to the falls and got lost in the timber along Frog Creek. Dora had a .22 rifle with her and once she was certain that they were lost and competely disoriented, she decided to stay put and fire her rifle from time to time through the night – which was the coldest night of that winter season. In the morning they were found by a searching party of about 50 local loggers who had been looking for them all night. "They suffered no injury except the fright of being lost," according to the newspaper. Staying put worked that time.[3]

If the strategy of moving on is used, lost persons tend either to veer in

[3] The story is told by Charlotte Mahaffy in *Coos River Echoes*, pp. 45-6. Also, see *Coos Bay Times*, Jan. 15, 1917, "Two Girls Lost."

one direction – thus "going in circles" as they say, or they may zig-zag, first one direction then back. That seems to be what Ondine did, endlessly walking away from the creek, then back toward it, always tending in a southerly direction – further and further from their house but nearer and nearer to Glenn Junction and houses. By the third evening she had probably got as far as the overgrown trace of the spur of an old logging road from the 1930's where she curled up under the huckleberry for the night. With the first morning light she continued walking, now wet and suffering from immersion foot syndrome, the result of three days and nights wearing her rubber boots, and no socks. It is a wonder she could walk at all, but an hour later she came out of the woods at the site of the old Glenn Creek School, almost directly across the river from the house of Wilma and Fred Lund.

As the story goes, Wilma thought something was wrong with her water when she got up that morning. She and Fred went outside and walked to the edge of their yard looking at their water pipes at 8 o'clock on the morning of December 27, the same moment that Ondine was looking at those same water pipes from the other side of the river. Wilma heard a faint cry of help, looked across the river, and there was Ondine. And just in time.

Wilma and Fred rushed across their cable swinging bridge, picked up the little girl, took her to their house, gave her a blanket and fed her some soup and when the Coast Guard helicopter took her away to the hospital, that was the last Wilma ever saw of Ondine. After all the excitement, it turned out that there was nothing wrong with Wilma's water, or so the story goes.

Ondine marveled at how mixed up the newspaper accounts were in many of their details. Those accounts did not reflect the fact that she had got out of the woods completely on her own – she was on the edge of the East Fork, looking at Wilma's house when Wilma saw her.[4] If she hadn't got that far, she would probably have died in the woods that day because it was a cold rain and nobody was going up there.

But what of the truncated search and rescue effort? When to stop is always a judgement call, but this one does appear to have ended a bit soon. There is now an international statistical database that can be used as a guide, much of it organized and summarized by Robert J. Koester in his handbook, *Lost Person Behavior* (2008).[5] It reflects that

---

[4] See especially the article by Vince Kohler, *The World*, December 27, 1977, p. 1. It gives a confusing account which indicates that Wilma had found her in the woods while working on her water supply intake.

[5] Koester used data from 16,863 searches from the International Search and Rescue Incident Database (ISRID) as the basis for the statistics in his book. He divided the searches among various subject categories, one of which is "Child, 7 to 9." There were 194 cases in that category. See Koester: *Lost Person Behavior, A Search and Rescue Guide on Where to Look – for Land, Air and Water*, pp. 139-143. Charlottesville, VA, 2008.

a child 7 to 9 years old, lost in wilderness, stands a 98% chance of survival if found within 24 hours. Between 24 and 48 hours, survival chances falls to 69%, and over 48 hours it drops to 33%. If Sheriff Les Miller had access to any such statistical information (which he probably did not) one would suppose that the search would have continued through Christmas day or longer. However, the statistics show that by 72 hours, the chance of finding the child alive falls to zero. By the time Ondine emerged from the woods it had been 66 hours. She beat the odds, big time.

By contrast, on October 22, 1989, about 15 miles northwest of Chemult, Oregon, a ten-year-old boy named Nathan Madsen disappeared while riding his pony from his parents' cattle roundup to their base camp. 200 to 300 searchers continued an intense search for more than 30 days, without finding him. His horse turned up grazing in a meadow on November 19, and the search finally ended November 22, with a $15,000 reward remaining on deposit at the Key Bank[6]. The following spring the search was resumed but his remains weren't found until July 21, 1990, almost exactly 9 months after he disappeared. He apparently became disoriented much the same as Ondine had. His remains were found about a mile from where he was last seen alive.

My son-in-law Larry Otten was one of the searchers. He was a logging engineer with International Paper, the same company the boy's father worked for. The company sent 50 volunteers and a leased helicopter.[7] About 26 square miles were designated as the search area, divided into three sectors. The IP volunteers were given the north sector, west of the Little Deschutes River. Larry and the other IP engineers were placed in charge of the intense grid searches of their respective subsections, but nothing was found. He said that it was well organized, and well financed with a commissary where gloves, coats, socks, and food was available to any searchers in need. There was intense interest, fanned by the *Eugene Register Guard*, the *Bend Bulletin*, and other newspapers keeping it on the front page with large color photos for days, as well as local TV news coverage.

At the same time as the Nathan Madsen search, Joe Mills, a 25-year-old hunter, was lost in the Cascades east of Oakridge. Aside from his own family there was very little effort to find him. Just exactly what it takes to keep a search going has a lot to do with the stirring of emotions – the boy on his pony, the romance of the cowboy lifestyle – and a newspaper eager to keep the interest and the emotions flowing. That didn't happen for Joe Mills, and it didn't happen for Ondine Eaton, but the long and intensive search for 10-year-old

[6] *Eugene Register Guard*, November 18 and November 22, 1989.

[7] Ibid, October 25, 1989.

Nathan Madsen did him no good. Joe Mills, with a very short search, ended with a similar fate. And Ondine Eaton pulled through on her own, with no help at all from the search. If one is lost, unprepared, in the woods what is needed is a happy coincidence and a lot of luck. The chance of that happy coincidence occurring in time is much better if there is a big, well organized search party looking for the victim, but there is no guarantee.

Wilma Lund: 1915-2010

Wilma Lund's swinging bridge. Ondine appeared on this side of the bridge, saw Wilma and Fred and cried out. Wilma ran across the bridge and brought her back to the house.

Chapter 8

John Clarence Fish, Jr., and Fog, February 4, 1984

There is the fog of meteorology, and there is the fog of neurology. Either can be disorienting and when combined they are frequently fatal. Take the case of John Clarence Fish, Jr, scion of a pioneer family of the Coquille River who grew up loving the outdoors, hunting and fishing through his youth. Born on February 4, 1919, he had served with the rest of the Greatest Generation in WW II and rose to the rank of Chief Master Sergeant in the Air Force before he retired with 23 years service. In retirement he returned to Coos County where he engaged on a second career, keeping fit so he could continue to enjoy his outdoor life. He was a handsome man, a quite dapper 6 feet, 175 pounds, with grey hair under the driving cap he often wore. Home in Coos County, he could now relive his youth every day.

This long deserved happy life ended quite suddenly when in his late 50's a stroke laid him out for a time, with a slow recovery. The kind of stroke he suffered was a "non-dominant side" stroke which, according to his daughter, causes disturbances in judgement, orientation, and emotional stability.[1]

[1] Communication with Kathleen Moore, July 22, 2010.

There was also partial paralysis of his left arm and leg, but within a year only mild efects of those symptoms remained.. On February 4, 1984, he was looking forward to a 65th birthday dinner with his sister Joyce McCollum at her house on the West Fork, about three miles above Allegany. It turned out later that he wasn't feeling very well that night, and didn't eat much before he left her place at 7:30 p.m. to return home to North Bend. The sun went down at 5:30 that day, and so it was dark and he was heading downriver into one of those heavy West Coast fogs that sometimes show up in February.

I wouldn't have known any of this but I was active in the local theater with his step-daughter Candace Dickey née Pressnall. Cande had recently played the title role in a musty old historical play I had directed, *Queen Christina*, by the Swedish playwright August Strindberg. She not only did a convincing job of acting the part of the queen, Cande even looked like the queen when compared with a famous portrait of her. The playwright Strindberg had always been fascinated with coincidence and hints of the occult and Cande had the chills run down her spine when she removed the back of on old picture frame among the junk in her garage. The cardboard back of the picture frame had been cut from the lobby card of the 1933 movie *Queen Christina*, in which Greta Garbo played the same character Cande was playing at that moment. It was only one of several extraordinary events during the production of the play, a little after the end of which a very seriously concerned Cande came to me with a problem.

I was already working on another play, and was quite unaware of anything else that was going on. Cande told me that her stepfather had been missing for almost a week (Actually, John Clarence Fish, Jr. was her only father, as her mother had married him when Cande was only two years old). She told me that they were now certain that he was up in the Elliott State Forest, lost or dead, and asked if I could get Bob Mahaffy to help and the two of us join in the search. She said that they had gone to the police for help but were told they had to wait forty-eight hours. By that time it had actually been four days, and now it was five. It appeared that nobody was looking except the family. In desperation, Cande's sister called two churches she had contact with and organized a search for Saturday, February 11. Cande was very visibly upset about it and wanted Bob and me to help.

Her sister Kathleen Moore had arrived from her home in Bend and she and Cande had gone to the police, the TV station, and the newspaper with information concerning the car, what her stepfather was wearing, and where he had last been seen. All the publicity came from the family, who were told that the Coast Guard was engaged in the search, but they later found no evidence that anything had been done other that what they, themselves, had initiated. With publicity, there was hope that someone who had seen him or the car would come forward. On Thursday, February 9, it was on the local TV news, and on page three *The World* newspaper at last had an article with the heading, "Searchers fail to locate man."

And who were the searchers? First it was Cande and her boyfriend, then it was more of the family, and they had spent the days searching along the river for any indication of a car having gone into it, to no avail.[2] After the TV news of the disappearance of Mr. Fish, his family received a call from a man who lived at the "Devils's Elbow" on the West Fork, about two miles above Allegany. He had talked to Mr. Fish on that fateful night. He remembered that indeed a man of that description had pulled into the driveway asking for

---

[2] There had been cars go into the lower part of the river in the past, and had not been found for weeks. For example, on the night of July 30, 1958, Allan and Anita Erickson and their their baby boy disappeared after leaving relatives in Cooston. An intensive, two-week search along all routes from Cooston to their home in North Bend revealed no clues. Eight weeks later, a fisherman noticed an oil slick in the river about a half mile above the Catching Slough bridge on the Coos River highway. The Sheriff's office brought in a scuba diver and sure enough, it was the Ericson car with the bodies of the three of them together in the back seat. See *The World*, September 24, 1958 for a full account.

directions, but there was no indication that he was in any distress – all seemed quite normal and he told Mr. Fish that he had merely turned the wrong way onto the driveway. There had been no trace of the car downriver and this new information left only upstream to search, into the headwaters, into the Elliott.

I hadn't seen the article on page three of the newspaper, and we didn't receive TV where we lived.. The first I heard that anybody was lost was when Cande appoached me about helping. She said that there would be an active search in the Elliott starting in the morning. In the morning I thought, "There are 92,000 acres up there with 450 miles of logging roads. I don't know where to look or what to do – somebody who understands that maze should be in charge. I called Jerry Phillips, manager of the Elliott, and asked if his people were active in the search. I got the impression that he hadn't heard that there was a search going on in the Elliott. I pleaded with him to get his people involved, as the family seemed pretty certain that the Elliott is where he must be and they needed help.

I rode with Bob Mahaffy in his pickup. On the 2000 line a little beyond Trail Butte we stopped to talk to somebody in another pickup who had a CB radio. While we were talking a call came over the CB that somebody had found the car. They told where it was and we drove to it. About a mile from Trail Butte a spur logging road goes off to the right to the head of Schumacher Creek, the 2080. A few hundred yards down it, another spur goes off to the left and it ends at an old landing. On the far side of that landing a green 1977 Mercury Comet had run nose down over the edge. The doors were locked and no one was around. The radiator hoses had broken loose. It was the car that John Fish was driving when he left his sister's house six days earlier.

About the same time that Bob and I got there, Cande showed up. It was an emotional moment for her, especially after the car was opened and the clothes John had been wearing at his sister's were found, his brown slacks, tan shirt, and his jacket. He had apparently slept in the car that first night, then changed into outdoor clothes for the hike out. It is hard to say whether, in the light of day with the fog cleared, that he knew where he was, but I suspect that he had a pretty good idea.

I don't think it was more than an hour after the car was found that it was reported over the CB radio that somebody had found the body lying in the road on the 2100 line. The body was found by Al Krentz, a forester from the Coos Bay office. I later asked Al why he went down that particular road and he answered that he had been told that someone was missing, perhaps in the Ellliott, and he would kind of keep an eye out for him. But the reason he went down the 2100 road at that time on that day was because it had been raining quite a bit over the past week and a couple of miles down the road there was a big culvert that tended to plug up. He was going down the 2100 road to check on that culvert.[3]

The 2100 was built after the Columbus Day Storm of 1962 to access the salvage of some of the blown-down timber. The road was pushed through about five miles, to the farm of Fred McCulloch near Templeton on Tenmile Lake. There had only been two or three timber sales for the salvage, and the road hadn't been used for several years.[4] It went steeply down to the headwaters of Johnson Creek, and most of it was impassable to vehicular traffic. Logic would place it very low on the list of roads someone would have turned onto if they were trying to get out of the Elliott. The road twists and turns from the 2000 line on the ridge down through several switchbacks, and partway down it was closed by a slide. Al Krentz decided to park his pickup and walk on to the bottom to check the culvert, since he was that close. A little before he got to the culvert, he was shocked by the sight of a body lying in the middle of the road.

John Fish had walked about three miles from his car, heading in a direction that would have led him to help if he could have continued. Beyond where his body was found, it was only another three miles to the McCulloch ranch near Templeton on Tenmile Lake. By coincidence, he was barely a mile from the site of the airplane that crashed in 1954 and laid there 23 years before it was found (see Chapter 5). If Al Krentz hadn't been concerned about a plugged culvert, it may have been a very long time before anybody would have found the body of John Fish.

So what brought John Fish to this very unlikely spot? We can't be sure, but at least part of his disorientation must have been the result of his stroke[5] When he left his sister's house, he entered what Sir Francis Drake had complained of back in 1579 when he visited the southern Oregon Coast, "most vile, thicke and stinking fogges."[6] The damage to the brain from stroke could make it hard enough to recognize what should have been familiar, and the fog removed all frame of reference, stroke or no stroke.

About a mile from his sister's place is a 180 degree turn in the road called the Devil's Elbow. It is a wide turn, and in line with the road as you drive toward Allegany is a driveway which runs steeply down to a small group of houses along the river at the bottom. When he got to the Devil's Elbow the fog obscured the turn and his headlights shone straight ahead onto the driveway

---

[3] Interview with Al Krentz, July 2, 2010.

[4] Interview with Jerry Phillips, July 7, 2010.

[5] Micro-strokes are recognized as one cause of vascular cognitive impairment (VCI), a form of dementia.

[6] From *The World Encompassed by Sir Francis Drake*, 1628, (entry for June 5, 1579).

and he drove right down it. Quite confused when he got to the bottom, he asked for help from Linda Edward's new neighbor.

John Fish drove back up the driveway and when he came to the county road again, again his headlights shone straight ahead, obscuring the fact that if he turned to the right he would go on downstream toward his destination.[7] Instead, he drove straight ahead into that tunnel of light, his mental map a full 180 degrees out of whack with reality. He went back upstream, past his sister's house and on another four miles to the end of the paved road where it forked. To the right the county road continues another four miles where it dead-ends; to the left is the beginning of the Trail Butte road, the 2000 line of the Elliott State Forest, and an infinite maze of dirt roads, many of which go nowhere at all.

In a sense, John Fish was following the pathway of a labyrinth as he continued into the tunnel of light made by his headlights. On and on he drove, up a steep and crooked hill, to the top near Trail Butte, and on for another mile, to a choice. An unused logging road entered the main road from the right, and John took that turn to the right. In a true labyrinth there would be no choice and so now he found himself in what could technically be called a maze – which the next day would make clear.

The next morning walking out, John's first decision point was whether to go left or right when he came to the 2000. Left was slightly uphill, back toward Trail Butte, but also toward Allegany, the way he had come the night before. In that direction it would only have been about four miles to the first house on the West Fork. Right was slightly downhill, but it led into an infinite maze of logging roads. People who are lost rarely go back the way they came and John Fish was no exception.[8] He chose to go to the right, into the maze. This maze, however, had clues. On a clear day there are places along that stretch of the 2000 where there is a panoramic view over the western part of the Elliott, and for ten miles west to the sand dunes and the Pacific Ocean. John probably got a glimpse of that view and instantly and unconsciously the map in his head could be altered to more nearly match the reality of the landscape. He probably knew then pretty well where he was, and the most direct route to where he wanted to be.

His next decision point came about three-quarters of a mile further along the 2000. Here he could go to the left, off the main road and onto a dirt track designated as the 2100. As unlikely as it seems, that dirt track was actually the shortest way out of the Elliott from where he was, although one doubts that

[7] Among the symptoms of dementia is a narrowing of the visual field, creating "tunnel vision." See *Lost Person Behavior*, p. 161.

[8] It is also interesting that a majority of lost persons go downhill rather than uphill (32% uphill; 52% downhill), and when given the choice, they turn right more often than left. (see *Lost Person Behavior*, pp. 61, 68, 79).

he could have known it at the time. But he presumably thought it was going in the direction of home and so he made that fateful turn. He had left his jacket in the car, and had lost one of his shoes by the time he stopped. His daughters believe that he probably became exhausted and laid down, and in the cold succumbed to hypothermia. His strength had given out after only two more miles of what would have had to be a five mile hike to get out of the Elliot that way.

There seems no doubt that John Fish's stroke was the operative factor in his having got turned around and lost. Fog was the contributing cause. In his classic handbook, *Lost Person Behavior*, Robert J. Koester took 16,863 searches from the International Search and Rescue Incident Database (ISRD) and organized them into 34 subject catagories. The category that most nearly approximated the kind of symptoms John Fish was experiencing was the one Koester called "Dementia." For purposes of analysis, Koester found 1051 such cases in the ISRD and John Fish exhibited several of the hallmark behaviors the night that he got lost. For example, Koester says that lost persons in this catagory "go until they get stuck," and "they appear to lack the ability to turn around." We certainly saw that in the behavior of John Fish on the night of February 4, 1984. "In a short conversation with the subject one may not detect anything unusual," Koester says. That could well explain why the neighbor did not see the need to give more expicit assistance when he talked to Fish that night.

I think that his actions the next day, the decisions that got him down in the bottom of the 2100 line at the headwaters of Johnson Creek, may have had a more rational basis. I think he may have known very well what he was doing at that time – he was going in a direction that would have got him out of there. According to his daughter, the autopsy showed no evidence of further strokes. The cause of death was hypothermia.[9] He had merely gone as far as he could go.

1977 Mercury Comet

[9] It is also possible that Mr. Fish could have experienced a transient ischemic atttack (TIA) – a micro-stroke or series of them – on the evening of February 4. Symptoms may include cognition disturbance resulting in temporary confusion or uncertainty, symptoms which are consistent with his actions on that evening. Such transient attacks may very well not show up as part of the autopsy but in this case could have been the major contributing cause, along with the fog, of his getting lost in so unlikely a place.

Chapter 9

Vera Edwards, Diane Wallis, and Slides

After the Coos River drainage had been criss-crossed with logging roads into every part, it was far less likely that there would ever again be a need for searches such as the one for the Army flyers in 1945 or the elk hunter Clayton Carroll in 1950. Because the roads went almost everywhere within the forests and had their own hazards, in future there would be fewer searches and more rescues. Stories abound concerning pickups that went off the roads that run along the ridges, sometimes several hundred feet straight down, and drivers and passengers sometimes had to be rescued. But the roads that go up the canyons also pose dangers of their own, mostly from rock slides. Cars can crash into slides, and sometimes do, or run off the road where it slides into the river. But the most dramatic of all accidents on these roads are when the slides crash into the car, and cover it up. That happened twice in recent years.

On Tuesday morning, March 5, 1991, some of the people going to work from above our place at milepost 18 on the East Fork of the Millicoma noticed that small rocks were coming down in slow spurts from above the road. It continued through the day, and late that afternoon Arlo Long stopped at our house and said that it looked like a slide was coming – the side of the hill above the road a mile upstream was slowly moving as he passed. A little later my wife Hilda was on the phone when Diane Wallis's little Nissan Sentra passed the house. She was coming home from her job at Payless in Coos Bay. Two or three minutes after her car went by, the telephone suddenly went dead. An hour or so later – it was quite dark – Jim Kellum and Pete Howerton rushed to our house to use the telephone. They said there was a big slide a mile up the road with a car under it.

Afterwards, we heard the story of how Pete Howerton was walking on top of the slide (something that nobody ever does). He was looking for his daughter who was due home and because the telephones were out, she hadn't called. Jim Kellum and Kip Maine drove up and joined him on top of the slide. Over the roar of the river Jim heard what turned out to be the muffled sound of a car horn coming through the mud under his feet. They scrambled around to find how the sound was getting to them and discovered the broken back window of the car hidden under the limbs and boughs of the trees that had come down in the slide. The rest of the car was coverd in rocks and mud, no telling how deep. They were surprised that it was not Pete's daughter, but Diane Wallis who was stuck under the slide. She heard Jim's loud voice above her and with her right hand, all she could move, was honking the horn from under the mud. That evening at the hospital she told *The World* reporter, "Jim kept talking really loud. That was the only thing that kept me going."

Logs bucked out of the way to give access to the door; dirt and mud dug out so that the door could be forced open, releasing Diane from her trap.

A boulder the size of a house, weighing two or three hundred tons, had landed inches from the car, demolished the road, and rolled over into the river where it will remain forever more. The car was buried in mud and debris with two uprooted Douglas fir trees on top of that, branches concealing the back of the car which was sticking out of the slide toward the river. Diane was immobilized inside, with her head pinned between the headrest of her seat and the roof of the car and water dripping into her left ear for two and one-half hours. All she could move was her lower right arm, and with it was able to reach the horn. Mercifully and inexplicitly the dome light remained on, a godsend for Diane who tended to be afraid of the dark. At one point she heard the diesel engine of her husband's pickup approach but it stopped, turned around, and went back up the river. For her, that was the point of utter despair, but then later came the welcome voice of Jim Kellum and she knew that it would be OK.

Because our phone was dead, Jim and Pete went a half mile further down to Ben Carter's house and called 911. They grabbed a chain saw, an ax, and a shovel and along with Ben, they hurried back. I was at work at the time, but our friend Arlo Long and our son Oliver happened to be at the house and they grabbed a shovel and went to help. In the darkness, logs on the top of the car were bucked out of the way with the chain saw, and enough of the dirt and mud was dug out to expose the driver's door. Together, with all their strength, they were able to force it partly open.[1]

Drivers seat with roof against the head rest.

Oliver said that he dreaded to look inside, expecting blood and gore, but Diane was miraculously unhurt. Once freed from her prison the dome light, for whatever reason, went out. She was carried across the mass of mud and rocks, arriving on the road just as the ambulance arrived from town, and at that same moment her husband Bob arrived from their home four miles above the slide. At the hospital she was treated with a cut on her right hand and released. Later she had an ear infection from the water that had dripped into it for two and a half hours, the only ear infection she ever had in her life. She told me that sometimes, when it rains hard, she re-lives some of it. All in all, it was a very lucky ending to something that could just as easily have been a horrible tragedy.

Two years earlier there was a tragedy on the South Fork. At a little after eleven o'clock on Sunday morning, January 15, 1989, someone reported to Weyerhaeuser that there was a slide partially closing their road at mile 9.5 above Dellwood. Although a private road, the guard at the Dellwood gate could let vehicles proceed if they had what he considered a legitimate reason for being there. On that Sunday morning, at least two cars had gained entry.[2]

The slide was on the south side of the river at a place were the road had been blasted out of a quarter-mile of sheer rock bluff, a rather spectacular spot. By the time the Weyerhaeuser road crew arrived about 1 pm, the slide was 60 feet high and 80 feet wide, completely blocking the road. While surveying the situation, they discovered the front end of a car, a vintage 1955 Chevy sedan, which had been crushed under a 20-foot-wide boulder. Someone was alive inside, a child.

---

[1] Details are from Arlo Long's daily diary for March 5, 1991, *the World*, March 6 and 7, 1991, and conversation with Diane Wallis. There were apparently six neighbors in the rescue party: Pete Howerton, Jim Kellum, and Kip Maine initially, then Ben Carter, Arlo Long and Oliver Youst.

[2] The account that follows is from my conversations with Ralph Sweet, who was present at the rescue and from two articles in *The World* newspaper by staff writer John Griffith: "Boulder crushes car, kills CB woman," Jan. 16 and "Girl's spunk aids in rescue," Jan 17, 1989.

The driver was Michael Edwards, 27 (a grandson of the pioneer steamboat captain Charles E. Edwards who operated the sternwheeler *Alert* on Coos River from the 1890's). It took him an hour, but he was able to extricate himself from the wreck without injury. There had been a rare mantle of snow on Blue Ridge that morning, visible from Coos Bay, and according to *The World* newspaper the family had been heading up the South Fork to see the snow. It was an odd route to take if that was the objective, but that was the story.

Michael's wife Vera, 24, sitting on the passenger side, had been instantly crushed to death when the boulder came down on the car. Their six-year-old daughter Tasha was pinned, immobile, in the back seat. Only her head was accessible through a six-inch gap.

By a peculiar coincidence of the type we find in many of these stories, a Coos County Search and Rescue team was on a training exercise in the area. Together with the Weyerhaeuser personel, they worked for about an hour, using a car jack and other tools to pry open a passageway that permitted emergency medical technician Mike Brown of Bay Cities Ambulance to crawl inside to be closer to the girl. The only part of her body that he could see or touch was her head. He stayed with her, holding her head and doing what he could to comfort the spunky little girl during the next seven hours of a very dangerous rescue effort.

Sheriff deputies, State Police, search and-rescue-volunteers, and fire department personnel from Coos Bay, North Bend, Millington, and Coquille arrived to help. A Coast Guard helicopter was able to land a couple of miles upstream and stood by to evacuate the girl as soon as she was released from her trap. It remained on post until approaching darkness and deteriorating flying conditions forced it to return to base at North Bend Airport.

Weyerhaeuser furnished lights and generators, and work continued for another three hours after dark. It was a dangerous rescue, with rocks falling from time to time, and the front of the car at the edge of the cliff dropping to the river about 100 feet below. Hanging from ropes, chipping away at the boulder and using the hydraulic Jaws of Life to cut through parts of the car, the rescuers gradually and carefully worked their way toward the body of Vera Edwards and from there to her daughter Tasha pinned behind her. Mike Brown later told *The World*, "There wasn't a whole lot we could do for her at the time. We had no access to her arms or legs, only her head." When the roof pillars on each side of the windshield were being cut there was danger that they could have been helping to support the boulder. Brown said, "I thought about getting out, but I figured I had to be in there with that little girl."[3]

At last, seven hours into the rescue, after removing the roof and cutting away the dashboard and the front

---

[3] *The World*, January 17, 1989.

seat with pocket knives and scissors, Tasha was removed to the ambulance and with her father delivered to Bay Area Hospital. He was examined and released but Tasha suffered from a broken leg and other injuries. Two days later she was reported in "fair" condition. She had gone through a lot but in the words of Ondine Eaton, the subject of a previous chapter, "kids are tough."

According to her uncle, Tasha suffered from pain in her leg off and on through puberty "but seems to have outgrown it." She had a hard time growing up, and from the age of about 13 was "pretty much on her own."[4] Her father died a few years ago, leaving her with no close relatives. She has since married, however, and has a child of her own. We can only wish her the greatest happiness from now on. She deserves it.

The Chevy under the rock. Photo still from VHS video, courtesy Ralph Sweet.

Arrow points to the front end of the Chevy. Vera Edwards in the passenger seat was instantly crushed to death. 7-year-old Tasha was pinned in the back seat for 7 hours. Her father, Mike Edwards, escaped unhurt. Photo courtesy Ralph Sweet.

[4] From telephone interview with Darren A. Simmie, December 18, 2010.

Milespost 19, Coos River Highway (East Fork, Millicoma River), March 6, 1991

Arrow, left, points to the door of the Nissan Sentra, which had been dug out and forced open. The arrow, right, points to the 300 ton boulder that missed the car by inches as it smashed the roadway and rolled into the river. Photo by Mike Sweeney, staff photographer, *The World* newspaper. From *The World* newspaper collection, 5/6/1991, CCHMM.

Chapter 10

Harold and Erma Ott, and Fog: January 22, 1994

With our aging population we may expect more cases of the bewildered to be lost in our wilderness. But it doesn't have to be a wilderness, and you don't have to be bewildered. Healthy and rational people can be lost only a few feet from the public roads.

The case of Harold and Erma Ott provide a really sad example. Harold understood fog, probably better than almost anybody. He was captain of the riverboat *Welcome* for more years than we care to remember. The boat was owned by his father, Jesse Ott, who was captain before him and captain of the steamboats that came before the *Welcome*. No one can know how many trips when the fog was so thick you couldn't see across the river, but my sister rode that boat from Allegany to the Coos River High School all four years, and there was never a trip that didn't make it irrespective of weather: freshet, flood, or fog. In the thickest fogs going into town in the mornings, Harold would place his pocketwatch alongside the compass and navigate by dead reckoning. He would go so many minutes at one heading, then change heading and go so many minutes in that direction, until he got to the Coos Bay Mutual Creamery where the milk cans were unloaded. The *Favorite,* coming down the South Fork from Delwood, would wait in the fog at the forks of the river until Harold came along, then would follow him into town. He was infallible.

Harold Ott

Erma Ott

Harold and Erma lived at Allegany in the house they built during the Great Depression. Their son Roger lived a few hundred feet below them on the West Fork, and Roger's son Dale lived with his wife and family on his great-grandfather's place above the road up from the store a little bit. The family had been here for as long as there was an Allegany. Harold once told me how the road finally opened all the way to town. The Hauser Construction Company had a rock quarry where they got the rocks to build the south jetty at the harbor entrance to Coos Bay. That job was finished in 1929 and it wasn't until the rock quarry closed that the Coos River Highway finally opened so that cars and trucks could drive to town. After that the river boats were less important, but they continued to run until December 1948. In 1951 the gravel road was paved.

The rock quarry was later made into a county park, named for the families who had originally owned the land: Rook and Higgins, Rook - Higgins County Park. In the days of the quarry, the barges were loaded with rock alongside what became the edge of the road, and there was no shoulder there. Other highways had white fog-lines on either side so that motorists could readily distinguish the roadbed from the shoulder and beyond. There were no such lines on the Coos River Highway, although residents (including myself) had repeatedly brought it to the attention of the Highway Department.

Harold had studied to be a theater organist – to play the Mighty Wurlitzers to accompany the silent movies in the picture palaces. He graduated in 1927, the year that The Jazz Singer, the first talky, came out and nationwide rendered 60,000 theater musicians technologically displaced, unemployed almost overnight. That did not diminish his love of movies, however. After the "big dish" satelite TV came out, he and Erma spent many of their evenings enjoying the old movies that were shown. Harold would stay up with his remote and punch out the commercials as he taped each and every one of those movies as he watched them so that he could watch the best of them again, and again. Sometimes he would give me one that he was particularly fond of, insisting that I would like it (which I invariably did). It was the perfect way to relive the happy days of his youth, remembering those wonderful old black-and-white films he loved so much.

Welcome and *Favorite* racing on the lower Coos River. *Welcome* ran to Allegany on the Millicoma River; *Favorite* ran to Dellwood on the South Fork, Coos River.

Sometimes he and Erma would like a little diversion. They would get into the late model Buick that he usually drove and they would ride into town and back, a 30-mile round trip that would take maybe an hour and a half. This was in pleasant contrast with the full day it used to take in their younger days to go to town and back on the river boats. I think they savored the luxury of it, going to town so easily, and on a whim.

On the evening of January 22, 1994, Harold and Erma decided that they should to go to town and get somthing for a salad Erma was making for a potluck they would be attending the next day. It was Saturday night and it was something to do. They got into in the car and a half-hour later they were pulling into the Safeway parking lot. They went in and shopped around for a little while, paid cash for a can of something (my wife thinks it was beets, but nobody remembers for sure what it was). It was 7:20 pm, according to the time printed on the sales slip.

On the way home they ran into spots of thick fog in places along the river and approaching Rook Higgins County Park it was so thick it was not possible to see across the road. Harold slowed to a crawl but even so, his right front wheel slipped off the pavement and onto the gravel shoulder of the road. He ran for about one hundred feet when the right front wheel dropped off the road, taking the car and its passengers with it, nose down into the river. Neither Harold nor Erma were hurt – it was a soft landing. Harold, by force of habit, pulled on the emergancy brake, shifted into park, and took the keys from the ignition. They opened the left rear door and got out onto the riverbank. But that was it. There was barely room for them on the bank, and it was fifteen vertical feet to the road above them and they couldn't climb up. The tide came in and the tide went out and they both died of exposure or drowning.[1] Harold was 82; Erma was 81. They had been together for 62 years.

I know that the fog was the problem that night. It was Saturday and my wife and I had been to the theater. Returning home there was fog, but as we approached Rook Higgins Park where the road and the river come together, it got thick, thicker than I ever remember it. I could not see the road and so I slowed down until further along, after the road pulled away from the river, the fog thinned out somewhat. But afterwards, we both often thought that Harold and Erma were probably on the edge of the river, only 15 feet from us as we passed in the fog that night.

Hauser Construction Company rock quarry about 1928. This is the present site of Rook-Higgins County Park. Harold and Erma's car went into the river about where the barge is being loaded (right center of picture).

The next day their son Roger and grandson Dale walked along the road, searching. At about 5 o'clock that afternoon they came to the Rook-Higgins Park and the place where the front wheel had got off the road. They followed the track, and then they saw the car, nose down in the water. Down below, on the river bank, they found the bodies. Dale gave a choked, surprised sob, saying "Grandpa."

[1] I am relying on my memory of discussion with Roger Ott after the funeral, and on *The World*, p. 1, January 24, 1994.

Later, I wrote a letter to the District Manager of the State Highway Department and another letter to State Senator Bill Bradbury telling of our past failed attempts to get fog lines painted on the Coos River Highway, and I told in the most vivid language I could of the tragic end of Harold and Erma Ott, the result of having no fog lines. A few days later, fog lines were being painted on the Coos River Highway, and for good measure they painted fog lines on the Powers Highway as well. That's what it took to get the attention of the State Highway Department. Roger told me that when they were taking care of effects at his parent's house, they discovered an unopened can of beets in the kitchen cupboard.

1993 Buick Century 4-door Sedan

Chapter 11

Firemen to the Rescue

Sharren Dalke was a working mother of six who lived on the West Fork about seven miles above Allegany with her husband John, at that time a long-haul truck driver, and two of their teen-aged sons. Another son was in the Marines and there were three grown daughters living away from home.[1] Sharren and John had bought the property, a subdivided part of the old John Hansen homestead, in the the late 1970's and for several years used the cabin for vacations.[2] Eventually, they were able to build a new house and move into it so that their children could grow up in a rural setting and attend the public schools in Coos Bay. They never regretted their decision because it worked out very well for the upbringing and education of their children, but the events of Thursday March 13, 1997, put the family to the severest kind of test.

Sharren was a dental technician, and she worked in North Bend for Dr. Dane Smith, oral surgeon. She drove a Volkswagon Beetle, very economical as a work car in her daily 50-mile round trip down a crooked road. One side benefit of going to town every day was that she could stop at Safeway to pick up a few groceries on the way home. The Safeway supermarket in Coos Bay had recently made VHS movies available, quite reasonably if they were returned in time to avoid the late charges. Sharren had movies at home that would be due and she put them in the car to return them on that fateful Thursday morning.

She didn't show up for work that day, and she didn't call in, something unusual for her. There may have been any one of a number of reasons – something came up at home, perhaps? However, by mid-morning her daughter Genevieve became aware that her mother hadn't arrived at work and she knew immediately that something was seriously wrong. Genevieve was dating John Saxton, a fireman with the North Bend Fire Department who was on duty at the time. She called to tell him that her mother hadn't shown up for work, and

---

[1] The children are Renée, Jody, Genevieve, Harley, Ryan and Robert.

[2] John Hansen was a German sailor who is said to have arrived in Coos Bay aboard a German sailing ship and homesteaded the 160 acres to the east (upriver) of the Bremer homestead described in Chapter 2. Hansen lived to a very old age, passing away in about 1950. His son Bill sold the property to Julius Swanson, who built a house on it and sold the timber to Smith and Jensen's gyppo sawmill. The property was then purchased by John and Ann Livingston and later subdivided into several small tracts, including the Dalke tract. (Interview with Al Lively, Jan. 19, 2011).

asked what she should do. As the day progressed and she could find no trace of where her mother had disappeared to, she continued to call John for support and advice.

Her brother Harley was at Marshfield High School. She went there to ask him if there had been anything special or unusual about their mother leaving for work that morning. He mentioned the movies, and Genevieve called Safeway to find out if they had been returned. The answer led them to believe that the movies had been returned, so it was assumed that Sharren had made it to town. The search effort lost its focus at that point, because if she made it to town, where could she be? Sharren told me that someone had cynically suggested, "Maybe she's having an affair someplace!"

But there was serious concern. Dr. Smith's office had notified the Coos County Sheriff's Office at 4 p.m., reporting that Sharren was missing. According to the paper, "Deputies looked, but failed to find her."[3] That was certainly not a physical search. She was not in the jail, and was not on any list at the Sheriff's Office. In those days a missing person report such as this one would probably have generated no further action, and the newspaper could truthfully (and ironically) say that the "deputies looked!" Searches for lost persons are invariably initiated by family and friends, but in this case the lost person was most fortunate in having professional first responders concerned about her.

As soon as John Saxton was off shift he called five other off-duty firemen and asked if they would help to find his girl-friend's mother. Quite outside the formal structure of Search-and-Rescue, the six firemen went on the search because searching for lost persons and saving them was what they did, and they were on this particular search because one of their number was associated with the family involved.[4] There remained the assumption that Sharren had made it to town, but at 6:15 Harley was home and received a call from Safeway telling him that his mother had three movies overdue as of 6 o'clock. There was now no doubt. She had never made it to town. She was in the river or off the side of the road someplace. Now the search could be focused.

Volunteer firefighter John Entgelmeier told me, "We didn't have anything better to do!" And so the six of them paired off in three pickups and headed out to Allegany, carefully checking the river and roadside along the way. They arrived at the Allegany store

---

[3] *The World*, March 14, 1997, p. 1.

[4] The six firemen who came to help was off-duty North Bend fireman John Saxton, off-duty Coos Bay fireman Mike Hurley, and North Bend volunteer firefighters Mike Johnson, John Entgelmeier, Hank Parrott, and Jeff Common. *The World*, March 14, 1997, p. 1.

a little before 8 p.m.[5] Genevieve had already looked for evidence of the car going off the road but although it was well after dark it called for one more look, this time with professionally trained eyes. By now it had been 13 hours since Sharren had left home, and so there was no time to lose. The six firemen went into action with the efficiency of a military operation. John Saxton remained in his pickup at the Allegany Store. He had the fire radio that could communicate with the fire department in North Bend. The other two pickups started up the West Fork looking for any signs of a car off the road or in the river. John Saxton had almost decided to call in boats to search the river between Allegany and town but probably not more than five minutes elapsed when they got a call from Mike Hurley: he and Jeff Common had found the car. They hadn't yet found Sharren, but they found the car.

It was only a mile and a half from Allegany. On the hill that curves up to the Devils Elbow they noticed a broken branch from an alder tree. They stopped the pickup, got out and walked along the road but there was nothing to indicate that a car had gone off at that place, or that the broken branch had anything to do with it. As they walked along, they saw nothing unusual and were about to turn back to the pickup when, shining their flashlight into the dense foliage there appeared something shiny about 80 feet below them. "You could catch it if you were walking, you couldn't see it if you were driving," one of the searchers said later.[6] It turned out to be the hubcap of the VW Beetle, which was upside down and wedged against two trees a few feet above the level of the river. Hurley immediatly radioed Saxton at the Allegany Store, telling him they had found the car but had not yet found Sharren. Saxton relayed the information directly to the North Bend Fire Department over the fire radio and emergency vehicles and paramedics were immediatly dispatched. In minutes, Saxton and the others were at the site of the accident.

Sharren's VW Beetle upside down, 80 feet below the road. She was pinned under it by her head and left arm and shoulder for 14 hours.

There had been showers off and on through the previous night and througout the day. The road was wet, not

[5] *The World* March 14, 1997, stated that the firemen "began searching along the road about 6 p.m." March 17, *The World* stated that the "search team spotted the wreckage about 8 p.m." That makes sense, and fits with the story of the Safeway call on the overdue movies at 6:15 p.m.

[6] Fireman Mike Johnson quoted in *The World*, March 15, 1997, p. 6.

an unusual condition. Sharren had apparently hit a slick place, overcorrected and went down the bank. Her Volkswagon had the old-style lap belts which, if they weren't fastened very tightly, could allow considerable movement within them.[7] As the car rolled, the centrifugal force pulled her up through the lap belt and partly ejected her through the sun roof. When the car came to rest on its top, she was pinned face down with one arm and her head between the roof of the car and the ground. By the time they were able to get her out, she had been pinned there for 14 hours with her face in the dirt and the weight of the VW on top of her. Fireman Mike Johnson was quoted as saying, "I don't even know where she got her air. She survived a hell of an ordeal."

The immediate challenge was how to get her out. A VW Beetle weighs about 1900 pounds. Being jammed against the two trees, and the steep slope of the bank, made it impossible for the six of the men together to lift it off of her. They had straps and ropes in their pickups to secure the car from moving, and one of the men used his Leatherman knife to attempt digging under her head in an effort to relieve some of the weight. They worked frantically until the first of the emergency vehicles arrived from Coos Bay Fire Department. It had air bags which were used to lift the car so that Sharren could be taken out from under it. She was placed in a stretcher, and manhandled up the almost vertcal 80-foot bank to the road.[8] The firemen knew what they were doing, it was not their first emergency, but connected as he was with the family John Saxton found this one to be emotional. He told me that Sharren wouldn't let go of his hand.

By the time they got her to the ambulance her head and chest began to swell due to the weight having been lifted from her head and upper body, "Then," John Entgelmeier told me, "you have another problem." The paramedic with Bay Cities Ambulance was Ivan Hultin, and he knew what to do. She was taken to Bay Area Hospital where on Friday she was listed with head and neck injuries in "serious condition." By Saturday her neck and upper body swelled dangerously – a symptom of crush syndrome – and she was listed as "critical," with a tube in her throat so she could breath.[9] John Saxton told me, "She

---

[7] The Oregon seat belt law was passed by a vote of the people after an initiative placed the measure on the ballot in 1989, the only state to pass such a law by vote of the people.

[8] Bob Hughes lived about two miles upstream and he had a drift boat. He knew the river, and he knew his drift boats – he was a licenced guide. He thought that if there would be a problem getting her up the bank to the road, he could take her in the boat to where the river met the road again. He arrived on scene with his driftboat, but it wasn't needed. One time he took me and my son Oliver down several miles of the West Fork, the only time I had ridden in a drift boat – a very memorable event.

[9] Crush syndrome is a serious medical condition that occurs after release of crushing pressure from skeletal muscle.

died once in the hospital." On Monday, while remaining in intensive care, her condition was reported as "fair."[10] Fireman Mike Johnson said, "She's a very strong woman to to have made it that long." All the hair and skin were gone from one side of her head where she had clawed herself while attempting to dig out with her one free hand. The fingernails and skin of her fingers on that hand were gone.

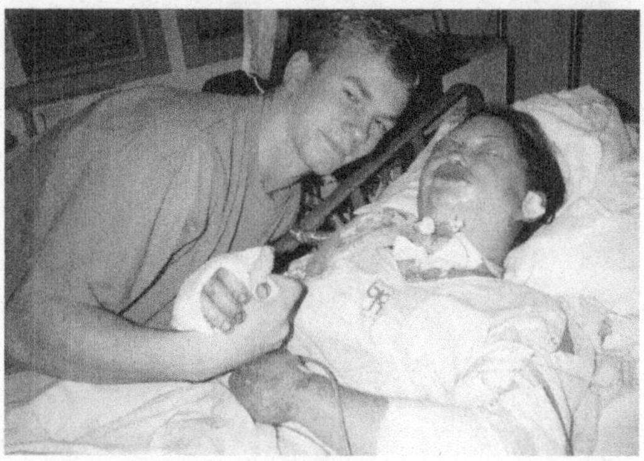

Sharren in the hospital with her son Harley, showing the extreme swelling, the result of crush syndrome after having the 1900 pound VW on her head for 14 hours.

It is sometimes astonishing, how long a person can live in such circumstances. Earl Stillman, 78, of Gold Beach went to sleep and ran off of highway 101. He lay pinned under his car for two weeks less than 100 feet from the highway before he died of dehydration and exposure. He kept a diary on the back of a road map. One entry read, "I need water most of all. With rain falling most of the time, I can get little by any device. I can hardly move without agony." He had lived two weeks and had been dead about five days when at last a passerby from Crescent City noticed the vehicle.[11]

But for the sharp eyes and strict attention to detail of firemen Mike Hurley and Jeff Common, Sharren Dalke may have shared the same fate as the elderly Earl Stillman. In her case the searchers got out of their cars and walked along the road. They found the clues that they would never have seen from the car and in the case of Sharren Dalke, it probably saved her life.

It is hard to over-emphasize the debt that is owned to "first responders" such as firemen. This came vividly into focus on September 11, 2001 when 343 New York firemen were killed following the terrorist attack on the World Trade Center. Overnight, firemen became our national heroes. That tragedy was quite fresh in the public mind the following

[10] Medical condition terms can be confusing. Going from worst to better is "critical," followed by "serious," "fair," and "good." Sharren was listed as serious the first day, then to critical, and finally to fair on Monday.

[11] "Diary Tells Story of Aged Man's Wait for Rescuers," *Coos Bay Times*, Nov. 13, 1945, p.1.

year, on November 25, 2002, when three Coos Bay firemen rushed into a burning business at 340 South 2nd Street minutes before the roof collapsed, killing all three. One of those firemen was Jeffery Common, 30, who along with Mike Hurley had located the wrecked car of Sharren Dalke. These guys live to save lives, they risk their own to do it, and sometimes they make the ultimate sacrifice.

Sharren made a 20-minute tape recording for me, telling of the accident in her own words. I have transcribed it and present it below, verbatim It is an amazing document of survival and rescue.

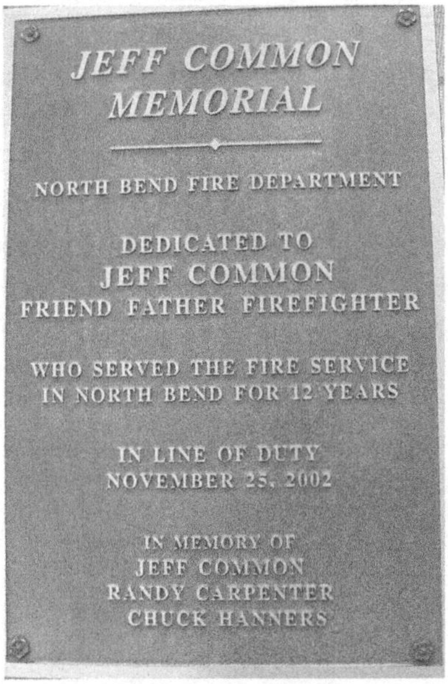

Sharren Dalke, her story

Good Morning, Lionel. Sharren Dalke, finally! I just thought I'd record this for you. Not very good at it, but here goes. The morning of the accident was Thursday, March 13, 1997. The weather was drizzly, rainy, grey outside, a usual morning in March for Oregon. So I was getting ready for work. I worked for Dr. Dana Smith, the oral surgeon in North Bend. I usually leave the house around seven o'clock. My son Harley was home and he was getting ready for school. I remember that morning because I had asked him if he would like a ride into town to school. Most of the kids jumped at that, because, you know, riding the bus – it's a long ride. So he had declined. He said, "No, not this morning." I said, "All right" Come to find out, he was chewing tobacco and he tells me later, "Do you know why I didn't ride with you mom?" I said, "No." He said, "Because I was chewing tobacco and I didn't want you to know." I said, "That chew of tobacco saved your life," because the first impact that hit the car would have been on the passenger's side. And it would surely have taken his life. So, I am blessed for that.

So I got ready, got in the car, I picked up my movies that I needed to return to Safeway, and into town I headed. Everything was going fine. I had a little blue Volkswagon bug. John and I were reconditioning it. It was cute. I can't tell you the year because I don't quite remember, probably was early nineties. Baby blue, with a moon roof. It was a cute little car. Good commuter car too.

Anyway, I was headed in and I had come around Devil's Elbow, headed down toward Gen and Cyn's house,[12] and I got about half way there and I hit something slick in the road and my car kind of spun toward the mountainside and then I counter corrected to the right. At that point my car went into a spin. I don't know, but I think it was diesel fuel. You know, there's lots of log trucks that go up and down. It's the only thing I can figure because there was nothing on the road for it to be so slippery. So it started spinning out of control. It spun off the side of the road. There was absolutely

---

[12] Geneva Thorpe and Cynthia Morris

nothing I could do. It started hitting trees. I think it went down about 80 feet. I just remember all the impact of the trees. And it came to rest upside down between a couple trees. I'm actually going to send you a picture of it.

At that point the moon roof had popped off. I had the old fashioned lap belts, that if you don't keep them completely tight, they will partially release you. That's what it did, it released me part way. Part of my head, probably from about the ears up and my left arm were out of that moon roof. And the car was coming down on me, crushing my head. All I could think about at that point was, "I'm gonna die. I'm going to die. What will my family do? What will my children do? What would my first grand-child?" which was Haley, Jody's daughter. I just, I couldn't believe it, I couldn't believe I was going to die. I felt I was too young, and I didn't want to go.

But as the car came down it was crushing, crushing, and I figured it was going to crush my head. And it stopped. Although there was a tremendous amount of weight, a tremendous amount of pain, it stopped, and there I lay. I had to wait a minute and take account of everything that was going on. I absolutely cannot move. I can't do anything. I couldn't even figure out where I was and what had happened, being jumbled around like that you can't put things into perspective. But, there was approximately fourteen hours to figure that out, unfortunately. I just kept waiting and my first thoughts were that a car's gonna come by, or a log truck would see me because they sit higher. After the first two log trucks drove by and nobody stopped, that's when I realized, nobody can see me. You know, in March is kind of springtime and everything was real green and real leafy, so they couldn't see me. They couldn't see anything down there. And I remember laying there. Of course I was conscious the whole time, and praying, and thinking about my family, and everything that had gone on, and how short life really is. Just in the blink of an eye, you can be gone.

On the other end of it, the reason I believe that nobody started looking for me earlier was, the office was closed that day. Dr. Smith's office was closed that day. But we always had somebody in there to answer the phones or make appointments and that was what I was supposed to do. And so they didn't realize. I believe Becky the bookkeeper had called a couple times and didn't get any answer, and came down and realized the office had never been opened, and so she started calling my daughter Genevieve and Renée. They couldn't get hold of John because he drove for Thomas and Sons and was out of town. All it took for Genevieve, she's my crusader I guess you would say, because as soon as she found out that nobody had found me she was on the phone to all her fireman friends. She was dating John Saxton at the time, so the ball got rolling pretty quick.

So they went and got Harley out of school and asked, "What was mom wearing, what was she doing, how did she feel when she left," and he said, "She was just going to work, and she'd taken the movies to Safeway." So they go, "Oh!" And they decided they had better call Safeway and see if I actually got to town because if the movies had already been turned in, I had made it into town somewhere. So, it's funny what people hear – what you say and what people hear. My daughter called Safeway and asked if there are any Safeway movies out for Sharren Dalke. What the girl heard, "Is there any late movies for Sharren Dalke," and the answer was "No." No late movies. So, that sent the whole search and rescue thing into a tailspin because now everybody thinks I made it to town. They told Harley to go on home, sit by the phone, maybe I'd call there, maybe somebody would call there. Harley went ahead and rode the bus home. At that point my daughter didn't believe it, so they're still looking.

And it's 6:15; Safeway calls because at 6 o'clock the movies are late. So at 6:15 Safeway calls and says, "You have late movies out; you have three late movies out that haven't been turned in." So everybody at that point went, "Oh my gosh! She never made into town. She's in the river or somewhere." So that's when the full-scale search came out. And I mean everybody. Debbie Fennel and Jim Fennel and Cyn and Gen and Bruce and Carole, the list could go on, Wendy, everybody up the river.[13] Everybody was out looking for me. Bob Hughes at one time – that's a little later, that was after they found me – Bob Hughes brought his drift boat up there. Because of where I was, how far down I was, they weren't sure they could get me up the bank. But that's kind of out of sequence, sorry.

So, they couldn't find me, and couldn't find me. I mean they were looking and the prayers were going up and all the churches in town, it was amazing. It was just amazing. We had become pretty well known because of our kids. They were so much into athletics and I worked for SWOYA[14] at the boys and girls club so we knew a lot of people, a lot of people were praying. So what had happened, I guess they were looking for oil slicks on the water because that's usually how they find people. And it was getting dark and they couldn't see and they were about to stop the search for the night when Jeff Common and Mike Hurley said, "We're going to go up one more time and kind of look."

And it was funny, there was just one broken branch, that's all there was for all that violence going over and off. And so they said, "Let's check it out." And I remember hearing that truck stop and then slamming the door and I went, "Thank you Jesus," because I knew they were gonna find me. They said they looked down there with their flashlight and got a reflection, and it had to have been one of my hubcaps. They said, "Well, we can't not check this out, even if it's a beer can or whatever." So I hear them coming down and they're going, "Sharren, is that you?" And of course I'm going, "Yes! Yes! It's me! It's me! It's me, but don't touch the car." I was so afraid if

---

[13] Cynthia and Bruce Morris, Geneva Thorpe, Carole Hughes, Wendy West.

[14] Southwestern Oregon Youth Activities

there was any pressure put on the car of course, it was going to kill me because it was so heavy on there, the pain was just so intense.

But, you do what you've got to do. You know, by this time I had tried to dig myself out. I had my right hand that I could use and I tried to dig the dirt out from around my head. It's funny what you'll do to survive. You'll just do anything. I had torn all my fingernails off and exposed all the bones in my right hand. It was like out of a horror movie. When I would flex my hand you could see all the mechanisms working. Not a pretty sight. But you know, you just do what you've got to do. I wanted to live. I wanted to survive.

And so the guys came in and that's when all the ambulance and fire trucks and all the rescuers came up. I can't tell you how much I appreciate those men. They are truly heroes. Firemen are heroes. This makes it a little touch[ing]. So, they got down there and they radioed that they found me. They had to secure this car in every aspect, in all four corners. They took me out I think towards the doorway because it had to be lifted all the same time when they put the air bags in. Nothing could shift, otherwise it would have killed me. All these men with all these things, tiny things that happened. One of the ambulance driver guys got this little tiny shovel trying to dig a spot for an air bag. They were just amazing. It was cold and wet and muddy. That's when Bob Hughes had put his boat in and had drifted up to that spot seeing if it would be easier to take me into the drift boat and then pull me out by the bridge. But I'll tell you what, after they put the air bags in and lifted it they were able to pull me out. I'll never forget that feeling. It was such a relief to have that off.

And there we go again. All these men, hand over hand, on a gurney, lifted me up the side of that hill and into that ambulance. I'll never forget it. So my trip into town, siren wailing – and I have a quirky little thing that Doctor – the ER doctor, the trauma doctor– his name is Tersigni.[15] He tells me, "Yeh. It was my birthday and I was out for my birthday dinner and I hear this siren and I'm hoping oh, that couldn't be for me! No trauma! As soon as I said that, my buzzer went off and there you were!" I sent him a birthday card for quite a few years and said I hope you have a good birthday this year.

So we get there and the room is full of all these people. [My husband] John had just driven into town and walked in there. My daughter Jody had come over from Roseburg. My son Robert was in the Marines. He didn't get there for about four days, they wouldn't release him. But all my children were gathered around. My father-in-law from California had come. It was just amazing. Not to say that everybody in Oregon, just about. We were a big wrestling family and I had prayers and flowers from Portland all the way down to Medford. It was amazing and I know that it was the

[15] Dr. Steven Tersigni, Vascular and Thoracic surgery

power of prayer that saved me. I get there and I'm in the ICU[16] and my husband is saying, "I don't want your mother to be left alone at all. There's somebody to be with her all the time." He just didn't feel safe. There was something in his gut that said, "I just don't want her left alone."

And so my daughter Genevieve was in there. I always like her to sing to me. She has this beautiful voice. So she was singing songs for me, probably Christian songs, as I know her. All of a sudden my throat closed, and I died. My daughter is running out of that room, screaming like a maniac. It so happened that Dr. Tersigni was still on the floor and they "emergency trached" me.[17] I had been without oxygen quite a bit, at least four and a half minutes, so they were not sure what was left. I remember barely waking up, I don't know how many days – couple days later. They had neuro-surgeons, they had all kinds of people in there and they were testing to see how much of me came back. Of course, as we know, most of it's here!

So, I had a tracheotomy and they brought me back and my family and my community surrounded me, but my thanks and my everything goes to those search and rescue guys. Some time later we had a great big party and I have some wonderful pictures somewhere of all of them and I was able to thank them, each and every one, in person. And they got to see how I turned out. That everything worked out and I'm living and alive and here I am now and I still have my six children and I have thirteen grandchildren, and I'm so happy to be alive.

If this isn't complete enough or if I've forgotten anything just call me and then we can go over it. Thank you so much for doing my story, Lionel God bless.

Lionel, after I was re-listening to that, I didn't tell you what I looked like. Because of the pressure of the car being on my head it almost tore my left ear off. I was so swollen you couldn't recognize me. My eyes were swollen shut and I couldn't hear out of one ear and couldn't see for about a week or two. Maybe it's more like two weeks. I looked like a monster. I lost half of my hair from the pressure of the car being on it. So it is quite amazing, as far as lasting effects, just scars. My ear came back just fine. I have a trach scar, and some scars from my right hand where I told you I dug out. But all in all, unless someone showed you those original pictures, you'd be amazed. I'll send you one of them and you can send them back after you're done. I'll send you the car and the picture of what I looked like. Maybe it would help with the story and if not, that's ok. But maybe the more information you can pull from them, the better it would be. All right, I'll quit talking now! bye bye!

---

[16] Intensive Care Unit

[17] Tracheotomy, an incision in the neck which allows a person to breathe without use of the nose or mouth. Among the oldest described surgical procedures.

After her full recovery, Sharren (in white) and her daughters Renée, Jody, and Genevieve.

Chapter 12

Hazel Lydia Chamblen and Fog, October 20, 2000

There are situations in which nothing would have helped. On November 6, 2000, hunters found a red 1990 Toyota Corolla station wagon parked on a landing in the Elliott near the 1000 line, on the divide between Glenn Creek and the West Fork. It was totally abondoned, several miles inside the Elliott. It turned out that it was the vehicle last seen on October 20 when 86-year-old Hazel Lydia Chamblen dropped off her grandson at the Winston High School at about 7 pm. That was it. She disappeared from the face of the earth. It is said that she was showing symptoms of mild memory loss at the time of her disappearance. Hazel Chamblen has never been found.[1]

How did her car get all the way from Winston to the 1000 line of the Elliott State Forest, a distance of about 75 miles, and no direct route? If we assume that dementia, as the term is used by Koester, was a factor, then the

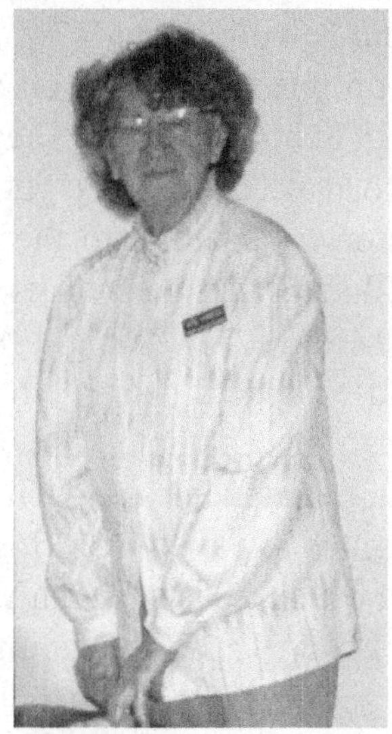

Hazel Chamblen, 1914-2000

hallmark behaviors that he listed were also present and we may be able to reconstruct a possible route that would have brought her red Toyota to that unlikely place.

First is the fog. October is the worst month for fog in the inland valleys of the Umpqua or the Willamette valleys. The fog that helped to confuse John Fish and Harold Ott was a sea-fog caused by the advection of moist air dropping onto the surface of the colder water of the Pacific Ocean, and it drifts over the land beginning usually in the afternoons. The fog that is so frequently prevalent in the valleys is caused by the advection of

[1] My thanks to Greg Kreimeyer, retired forester, for telling me about this case. For details, I am relying primarily on the Douglas County Sheriff's Office Missing-Endangered Person report, November 7, 2000, and the missing person website, briansdreams.com.

warmer air over the colder land surface, and it may be present at any time, day or night. The effect is the same: travelers can lose their bearings and get lost. And if, like John Fish or Hazel Chamblen, there have been symptoms of dementia, the fog might trigger a spell of hopeless confusion and bewilderment.

She dropped her grandson at the Winston High School at 7 pm. She lived on Reston Road at Tenmile. To get from either Winston or Tenmile to the 1000 line of the Elliott State Forest takes some doing, but she did it. If she got turned around and headed north toward Lookingglass, for example, she could easily have gone past Melrose to Umpqua and on to a series of paved BLM logging roads, one of which leads into Ash Valley above Loon Lake.[2] If she kept going, she could have got on the Douglas County Road which runs toward the head of Glenn Creek in Coos County. To get to where the car was found, she probably turned north on the mile-long Elliott State Forest road 1850, which would quickly bring her to the 1000 line (Elk Ridge Road). From there she ended up on a landing and disappeared.

It appears that Hazel Chamblen behaved exactly as might be expected in cases of dementia. One of the hallmark behaviors is that they "appear to lack the ability of turn around," and "they go until they get stuck." That is exactly what Hazel Chamblen did. It probably took a couple hours to drive through those BLM roads, if indeed that is the way she went. She more than likely followed the tunnel created by her headlights, the path of a labyrinth that brought her to the heart of darkness, where she disappeared forever.

It may safely be assumed that she got out of her car and wandered into the young forest that had been replanted after the logging of a few years earlier. There she died, no one knows how far from the landing. The trees will continue to grow until they are large enough for market, which may be 40 years from now. Until then, no one will have reason to go into that new, emerging forest where all trace of her remains will have reverted to nature.

Hazel Chamblen has the honor of being one of the only three persons who was swallowed up and left no trace in these woods. The others who vanished without a trace were Sgt. Robert T. Neal, the radio operator who bailed out of the Army C-46 that crashed on Weyerhaeuser land in November 1945 (see chapter 4), and Frank Serus who was lost in the splash of a log drive in 1951 (see chapter 3). Everyone else was found, either dead or alive.

1990 Toyota Corolla

[2] I am grateful to Jerry Philips for reminding me of the paved BLM roads as a possible route that Chamblen may have taken.

Map IX

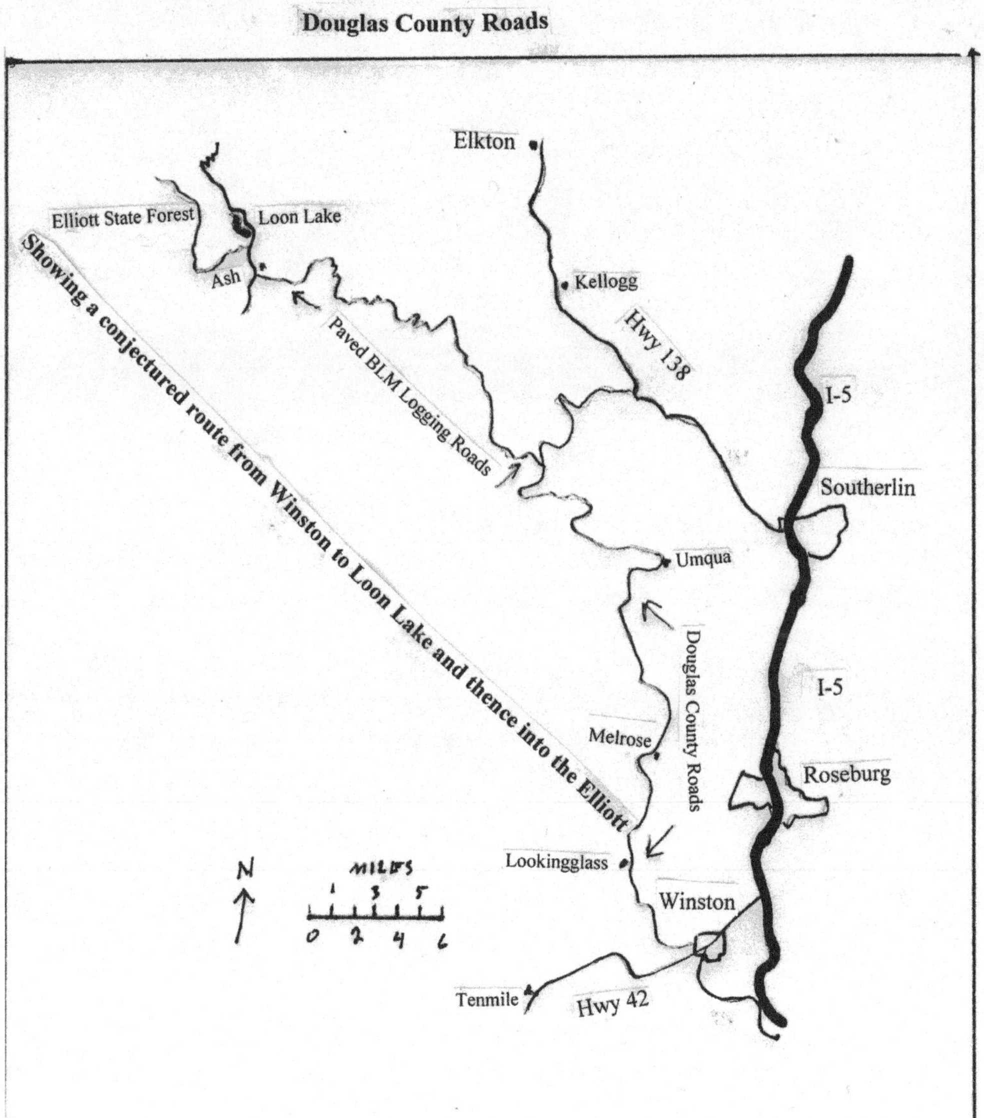
Douglas County Roads
Elkton
Elliott State Forest
Loon Lake
Ash
Kellogg
Hwy 138
Paved BLM Logging Roads
Showing a conjectured route from Winston to Loon Lake and thence into the Elliott
I-5
Southerlin
Umqua
Douglas County Roads
I-5
Melrose
Roseburg
Lookingglass
Winston
Tenmile
Hwy 42
N
MILES
1 3 5
0 2 4 6

Conclusion

As I was collecting information for these accounts, I was at first romantically infatuated with the stirring events of the pioneer period and assumed that nothing in modern times would ever equal the heroic deeds of the past. Very seldom would those old-timers admit to being "lost," and they invariably found their own way out. My friend Nathan Douthit has a better sense of historical perspective and he took issue with my contention. He convinced me that I needed to take a second look at it, and he was right.

Not even the most storied of the pioneers could surpass the feat of Roy Spires, who with his belt and spurs climbed a 250-foot Douglas fir at night to rescue the Army flyer: sitting on a limb, removing his climbing gear and putting it on the flyer, letting him down, pulling the climbing gear back up with his rope, putting the climbing gear back on and climbing down himself. That was in 1945. In 1950, we had Harold Noah's daring rescue of the lost elk hunter, bringing him down the full length of the West Fork in a rubber raft. In 1989 the emergency medical technician Mike Brown, at significant risk to his own life, held the head of seven-year-old Tasha Edwards for 7 hours during an exceptionally dangerous rescue operation. Those are rescues that certainly rank with the best that comes down to us from the nineteenth century. When the challenge comes to save a life, the altruism inherent in human nature may surface most unexpectedly.

Search and rescue is a serious, life and death undertaking. Of 27 lost persons who were subject of these stories, only eight of the comparatively uninjured flyers who bailed out of the C-46 could be said to have been alive and well when found. Ten of the total were found dead; three were never found, and are presumed dead. At least five were suffering from near starvation, four of whom had serious degrees of immersion foot syndrome (trenchfoot). Only one, seven-year-old Ondine Eaton made it out on her own with no help from the searchers, who had given up on her.

None of the lost persons in these stories backtracked. Without exception they charged straight on, into the unknown and in most cases getting further from help. There is the folk belief that following downstream will always get you to civilization and some of the lost persons in these stories tried it with near disastrous results. When the two lost hunters in 1871 decided that they had to go downstream from the headwaters of the South Fork of Coos River, they set themselves up for a hopeless venture that under the circumstances could not succeed. Some of the lost flyers in the Army C-46 that crashed in 1945 tried going down Lake Creek, which did them no good. That was also the apparent strategy of Clayton Carroll in 1950 on the West Fork. It did not work in any of those cases.

Another recurring theme is starvation and the problem of refeeding. The results of overfeeding may be mild, moderate, or severe – with death occurring in extreme cases. Cathcart understood the importance of very gradual refeeding of a starving man, but relatives of Beasley almost killed him by overfeeding after the rescue. Harold Noah understood the need for careful refeeding of the lost elk hunter in 1950 (Chapter 5), as did his cousin Wilma Lund when she took care of the lost girl in 1977 (Chapter 7). Both of these starvation victims were in the hospital very soon after their rescue, as was the flyer Walter Smoyer in 1945 (Chapter 4), and so one assumes that they all received responsible, clinically valid reintroduction of feeding. Beasley and Purdue had to rely on the nineteenth century folk methods of their rescuers, which were quite valid as it turned out.

When should a search be called off? That was as variable as the searches themselves. The State Police called off the search for the elk hunter Clayton Carroll at the end of the third day, saying that the area had been searched as "thoroughly as possible." The next day Harold Noah and Babe Page launched their own search and found the man six hours later. The seven-year-old Ondine Eaton walked out of the woods on her own after three days and nights, although the search had been called off after the first 24 hours. I suspect in hindsight that in both cases the lost persons would have been found by the original search party had they been more systematic in their approach. In the case of Clayton Carroll, a group of searchers going down the trail to Elkhorn Ranch may have found Carroll as certainly as Harold Noah had found him by going along the river. In the case of Ondine Eaton, a group of searchers would have only had to do a go through the one mile of brush and timber along the east side of Glenn Creek. They would have come across her; she was there, struggling along on her own.

And so, who is in charge of a search? In the case of Beasely and Perdue back in 1871 (Chapter 1) it was friends and neighbors. For Frank Bremer in 1938 (Chapter 2) the Coos Fire Patrol seems to have taken charge. In the splash dam drownings in 1945 and 1951 (chapter 3) it appears that the Irwin-Lyons Lumber Company took charge of the searches. The Army took charge of the search for the flyers who bailed out of the C-46 aircraft in 1945 (Chapter 4). The State Police in Reedsport were the prime agency in the search for the elk hunter Clayton Carroll in 1950 (Chapter 5). The Civil Air Patrol and the Douglas County Sheriff shared responsibility for the lost private airplane in 1954, although it had actually crashed in Coos County

(Chapter 6). The Coos County Sheriff took charge of the too-short search for the seven-year-old girl Ondine Eaton in 1977 (Chapter 7). From Chapter 9, it was relatives and friends who searched for John Fish in 1984 for several days before they were able to get the attention of the media and the Sheriff's Department. The Douglas County Sheriff took responsibility for listing Mrs. Chamblen as a missing person in 2000. It does not appear that Coos County was involved at all, and it did not appear in *The World* newspaper, although her car was found abandoned in the Coos County portion of the Elliott State Forest.

Traditionally in the United States it has been the sheriff and his deputies who are responsible for finding lost persons. In these stories it has been more complicated than that, and sometimes the sheriff has been involved and sometimes he hasn't. Most of the searches began with relatives and when they were unsuccessful, friends and neighbors came into the search. If they failed to find the missing person, law enforcement was called in, usually but not always the Sheriff's Department.

Happily, attitudes toward lost or missing persons have changed nationally during recent years, especiallly following the 2003 Homeland Security Presidential Directives. The changed approach in Oregon is reflected in the 2009 Edition of ORS Chapter 404, Search and Rescue. It stipulates specifically that the county sheriff is responsible for search-and-rescue activities within the county, and he is required to adopt a search-and-rescue plan. The plan contents are to include a detailed description of activities and circumstances that constitute search and rescue within the county, identification of volunteer organizations and procedures for contacting and requesting assistance from them.

Significantly, there are minimum standards for "qualified search and rescue volunteers," who must be registered with the sheriff. Registering with the sheriff has a real advantage. It holds the volunteer harmless for any tort claims that arise out of an alleged act or omission that occurs during the performance of search-and-rescue activities that are under the direction of the sheriff. Tort claims are matters that were never considered by the volunteers in the preceeding stories, but are no doubt of serious concern in these days.

There are new reporting requirements under the National Incident Management System. The State Office of Emergency Management assigns an incident number for each search and rescue, to be used for statistical and other purposes. If the search occurs in a multi-county area, then either one sheriff takes charge, or the counties "shall form a unified command." These are all positive changes.

According to my conversations with Sheriff Andy Jackson, Coos County has taken its responsibility very seriously.[1] He said that the Coos County plan was adopted by the Oregon State Sheriff's Association as the plan for the state.[2] Be that as it may, it does appear that current operations are far more focused, systematic, and professional than most of those described in this book. Sheriff Jackson informed me that now, as soon as a 911 call of a lost or missing person is received by the sheriff's dispatcher, the SAR coordinator is notified immediately and the ball starts rolling. There is no longer the delay of several days before the sheriff's office takes an interest. Once a search is underway it is supposed to follow established and proven procedures.

Better organization and more professional procedures help immensely with the odds, but cannot guarantee success. Success, now as in the past, so often depends upon the knowledge, audacity, determination, and bravery of a single individual – coupled with a lot of luck. In these stories there were many examples that illustrate the point.

Andrew Jackson, 1944-2011

---

[1] Sheriff Andy Jackson, b. 1944. He was with Coos County Sheriff's Office from January 1977, elected sheriff in 1998 and re-elected twice, serving through 2010 when he was elected to a 4-year term as County Commissioner. He died in office February 1, 2011.

[2] I assume that Sheriff Jackson was referring to a previous plan, not the plan required under ORS 404.120. Both he and SAR Coordinator Dan Stone were helpful in these final paragraphs.

Bibliographic Essay

I am not including a bibliography as such. My sources are in the text itself, or in the copious footnotes throughout the book. However, the following essay provides my best advice on what to read to learn about Coos River, its woods and its people, and about search and rescue in general. The published sources for Coos River history are fairly impressive. None of them address the subject of lost persons, or of search and rescue – that is the purpose of the present book. The sources discussed below are essential, however, to an understanding of early settlement of the area, and the exploitation of the timber resources up to the beginning of the 21st century. That is the setting in whch the lost persons of the present book found themselves.

For the early settlement of the Coos River drainage there are three essential books. The first is Charlotte Mahaffy's *Coos River Echoes: A Story of the Coos River Valley* (Interstate Publishers, Portland, OR 1965). This is a collection of stories, vignettes, and family histories that preserves much of the history of the settlers and settlement up through the mid-twentieth century. The second of this trilogy of books is Aileen Barker Rickard's *The Goulds of Elkhorn* (author, Cottage Grove, OR, 1982). This is a family history, centering on George Gould's settlement near the center of the "Big Burn" of the Coos Bay forest fire of 1868, raising his family there from 1885 to 1912. It is a most remarkable and poignant account of the phenomenon of the wilderness homestead. The third of these books is my own *Above the Falls: An Oral and Folk History of Upper Glenn Creek, Coos County, Oregon* (1992, 2nd Edition, Golden Falls Publishing, Coos Bay, OR 2003). Where the Goulds were settled on the upper West Fork of the Millicoma River, my book covers the wilderness homesteads on a tributary of the East Fork. Insofar as possible, I used tape recorded interviews with the people who lived there. I would like to think that it complements *The Gould's of Elkhorn* and supplements *Coos River Echoes.* An understanding of the settlers and settlement requires familiarity with all three books.

A personal history of the Weyerhaeuser Millicoma Forest is well documented by Arthur V. Smyth, who was Weyerhaeuser's first forester there. His *Millicoma: Biography of a Pacific Northwest Forest* (Forest History Society, Durham, North Carolina, 2000) is essential reading for anyone interested in what is often described as "some of the richest privately owned forestland in America." Another forester, Jerry Phillips – who was with the Elliott State Forest from 1956 to1989 – provided his personal history of the Elliot, *Caulked Shoes and Cheese Sandwiches: A Forester's History of Oregon's First State Forest "The Elliott" (1912-1996)* (author, Coos Bay, OR, 1997). It is exactly what the subtitle says it is, and it is essential reading for anyone interested in Oregon's first state forest. The logging, especially the "splash dam" river drives that were the principle means of transporting logs out of the Coos River drainage for the first 70 years of logging there is provided by Dow Beckham, who was river boss on the South Fork of Coos River during most of the years that splash dams were used there. *Swift Flows the River: Log Driving in Oregon* (Arago Books, Coos Bay, OR 1990) is the only book of its kind, a first hand account of splash dam log drives on the last river in the United States to transport logs in that manner.

There are three works of partly fictionalized memoir that are set in the Coos River drainage. First among them is Olive Barber's *The Lady and the Lumberjack* (New York, Thomas Y. Crowell Company, 1952). This is the exceptionally witty and insightful fictionalized memoir of a school teacher who married a logger who worked primarily on Coos River. Irma Lee Emmerson, with Jean Muir, wrote another fictionalized memoir, *The Woods Were Full of Men* (New York, David McKay Company, 1963). Irma was the cook at the Irwin and Lyons logging camp on Tioga Creek, on the upper reaches of the South Fork. Her descriptions of logging camp life, and her witty observations of the phenomenon of one woman in a camp with 60 men is well worth the telling. It was originally written as memoir, but her publisher insisted that it be fictionalized. The third of the fictionalized memoirs is A. E. Krewson's *Tiogas Pigs* (Portland, OR, Binfords & Mort Publishers, 1955). This is partly tall tale, but a good description of an interesting use of the very remote Tioga Creek myrtlewood bottoms, fattening hogs on the myrtle nuts and driving them back to the Coquille River for market. How much of it is based on true events and how much is tall tale, I don't know. But for anyone interested in the upper South Fork it is essential reading.

In compiling the information about lost persons and search and rescue I relied very heavily on contemporary accounts published in the newspapers. Microfilm copies of the daily *Coos Bay Times* and its successor *The World*, the *Southwestern Oregon Daily News*, and the weekly *Marshfield Sun* and *Empire Builder* are available at the Coos Bay City Library and were essential. Frequently, when I could compare the newspaper with the account by actual participants, the newspaper came up wanting. In some of them, the newspaper article bore very little relation to what the participants remembered. The strength of the newspaper articles is that they correctly documented the names and dates, and without them I could not have put the stories together at all.

There are two books and a few articles bearing on the general subject of lost persons and search and rescue that I relied on for context. First among them is Robert J. Koester's *Lost Person Behavior: A Search and Rescue Guide on Where to Look – for Land, Air and Water* (dbs Productions, Charlottesville, VA, 2008). This seems to be the current bible on the subject and I have referred to it frequently in the course of telling the stories. It is essential for an understanding of the way that lost persons, in various catagories, behave. Another influential book bearing on the subject is Laurence Gonzales' *Deep Survival: Who Lives, Who Dies, and Why* (New York, W. W. Norton & Company, 2005). Most of the cases described by Gonzales are of skiers, mountain climbers, and wilderness hikers – that sort of thing. These are not what I found in the Coos River drainage, but they are great stories of disaster and survival, well told About 10% of the book involves incidents on Mt. Hood near Portland.

An interesting journal article appeared in *Wilderness & Environmental Medicine*, June 2007, pp. 95-101, "Search is a Time-Critical Event: When Search and Rescue Missions May Become Futile" by Annette L. Adams, et. al. Its purpose was to validate a rule for duration of a search, after which a search and rescue (SAR) mission may be considered for termination. It was a retrospective study of all SAR missions in Oregon during the seven year period 1997 to 2003. The paper analysed the results of all the searches that lasted up to 51 hours, but they found that there was still a 57% chance of survival after 51 hours. Not too much help for the sheriff or family member trying to decide whether or not to call off a search, and I could not find in the article anything that would give me the basis of a rule for duration of a search. But it is a start. A more popular article appeared in *Backpacker*, 05 20 2010, p. 74 ff. "Utterly, Hoplessly, Truly Lost," by Jim Thornton. This is a very good summary and popularization of the findings especially of Koester in his "Lost Person Behavior." The psychology of the lost person is described convincingly and well.

Photo Credits

I have credited some of the photos with their captions, some I have not. The following list credits my source for each photo in the book. (CCH&MM: Coos County Historical and Maritime Museum)

Cover: Alice Allen
p. 3 Marshfield Sun
p. 12 CCH&MM #996-P17
p. 13 Lynne and Dennis Rice
p. 15 bottom rt: Lynne Rice
p. 15 bottom lft. CCH&MM #998-P135
p. 17 Curtis Irish
p. 20 author
p. 21 Alice Allen
p. 22 Alice Allen
p. 25 Melissa Green
p. 28 bottom lft: Adeline Larson
p. 28 top rt: Jack Slattery
p. 31 lft: author
p. 31 top rt: author
p. 33 CCH&MM #988-NA13
p. 36 author
p. 38 Coos Bay Harbor, CCH&MM
p. 39 Coos Bay Harbor, CCH&MM
p. 45 Coos Bay Harbor, CCH&MM
p. 47 bottom rt: Alice Allen
p. 47 top lft: Alice Allen
p. 50 Alice Allen
p. 51 Alice Allen
p. 53 Kathy Spires Larsen
p. 56 Alice Allen
p. 57 Jerry Baughman
p. 62 Lynne Rice
p. 65 Alice Allen
p. 68 bottom lft: Jerry Phillips
p. 68 top rt: Howard Lorman
p. 72 CCH&MM #955-1.11833
p. 73 CCH&MM #955-1-11822.1
p. 76 MHS Yearbook, 1952
p. 77 author
p. 79 author
p. 84 Howard Lorman
p. 91 Ondine Eaton
p. 96 bottom lft and rt: author
p. 102 Google images
p. 104 CCH&MM World Photos
p. 105 CCH&MM World Photos
p. 107 Ralph Sweet
p. 108 CCH&MM World Photos
p. 109 Roger Ott
p. 110 Jack Slattery
p. 111 Jack Slattery
p. 112 Google images
p. 115 Sharren Dalke
p. 117 Sharren Dalke
p. 118 Author
p. 119 Sharren Dalke (cropped)
p. 124 Sharren Dalke
p. 125 Douglas County Sheriff web site
p. 126 Google Images
p. 132 Coos County Sheriff web site

Index

Made in the USA
Coppell, TX
06 June 2023

17743512R10090